Environmental Epidemiology

Edited by Paul Wilkinson

Open University Press

Open University Press
McGraw-Hill Education
McGraw-Hill House
Shoppenhangers Road
Maidenhead
Berkshire
England
SL6 2QL

email: enquiries@openup.co.uk
world wide web: www.openup.co.uk

and Two Penn Plaza, New York, NY 10121-2289, USA

First published 2006

A catalogue record of this book is available from the British Library

ISBN-10: 0 335 21842 3
ISBN-13: 978 0 335 218 424

Library of Congress Cataloging-in-Publication Data
CIP data applied for

Typeset by RefineCatch Limited, Bungay, Suffolk
Printed in Poland by OZGraf S.A.
www.polskabook.pl

Contents

Section 6: Epidemiological evidence 173

Acknowledgements

Open University Press and the London School of Hygiene and Tropical Medicine have made every effort to obtain permission from copyright holders to reproduce material in this book and to acknowledge these sources correctly. Any omissions brought to our attention will be remedied in future editions

We would like to express our grateful thanks to the following copyright holders for granting permission to reproduce material in this book.

p. 52	Anderson HR, Ponce de Leon A, et al, 'Air pollution and daily mortality in London: 1987–92', BMJ, 1996, Vol 312, pp665–669, with permission from the BMJ Publishing Group.
p. 188	Aron, Joan L., Ph.D., and Jonathan A. Patz, M.D., M.P.H., eds. Ecosystem Change and Public Health: A Global Perspective. pp309, Fig. 10.3. © 2001 The John Hopkins University Press. Reprinted with permission of The John Hopkins University Press.
p. 131	Adapted from E Cifuentes, M Gomex, U Blumenthal et al, 'Risk Factors Risk factors for giardia intestinalis infection in agricultural villages practising wastewater irrigation in Mexico', American Journal of Tropical Medicine and Hygiene, 63(3), 388–392
pp. 63–4	Reprinted from THE LANCET, Vol 360, L Clancy, P Goodman, H Sinclair and DW Dockery, 'Effect of air-pollution control on death rates in Dublin, Ireland: an intervention study', 1210–1214, Copyright (2002), with permission from Elsevier.
p. 77	Darby S, et al, 'Radon in homes and risk of lung cancer', BMJ, 2005, Vol 330, pp223–228, reproduced with permission from the BMJ Publishing Group.
p. 189	Smith GD and Ebrahim S, What can mendelian randomisation tell us about modifiable behavioural and environmental exposures?, BMJ, 330:1076–1079, reproduced with permission from the BMJ Publishing Group.
p. 59	Dockery, DW and others, 'An Association between Air Pollution and Mortality in Six U.S. Cities', 329(24), 1753–1759. Copyright © 1993 Massachusetts Medical Society. All rights reserved.
p. 102	Reprinted from THE LANCET, vol 352, H Dolk, M Vrijheid, B Armstrong et al, 'Risk of congenital anomalies near hazardous-waste landfill sites in Europe: the EUROHAZCON study,' 423–427, Copyright (1998), with permission from Elsevier.
p. 186	Reprinted from THE LANCET, Vol 360, M Ezzati, AD Lopez, A Rodgers et al, 'Selected major risk factors and global and regional burden of disease,' 1347–1360, Copyright (2002), with permission from Elsevier.
p. 154	Gouveia N, Hajat S and Armstrong B, Socioeconomic differentials in the temperature-mortality relationship in Sao Paulo, Brazil,

	International Journal of Epidemiology, 2003, 32:390–397, by permission of Oxford University Press and International Epidemiological Association.
p. 155	Hajat S, Armstrong B, Gouveia N and Wilkinson P, 'Comparison of mortality displacement of heat-related deaths in Delhi, San Paulo and London,' 2004. Epidemiology, 15(4): S94.
p. 30	Reprinted from THE LANCET, vol 360, G Hoek, B Brunekreef, S Goldbohm et al, 'Association between mortality and indicators of traffic-related air pollution in the Netherlands', 1203–1209, Copyright (2002), with permission from Elsevier.
p. 158	Kovats RS, Hajat S and Wilkinson P, 'Contrasting patterns of mortality and hospital admissions during hot weather and heat waves in Greater London, UK,' 2004, Occupational Environmental Medicine. 61: 893–898. Reproduced with permission from the BMJ Publishing Group.
p. 169	© Crown Copyright 2005 Published by the Met Office, UK
p. 97	Reprinted from Moore KL, 'The Developing Human: Clinically Orientated Embryology', 6th Edition © 1998 Elsevier Inc, with permission from Elsevier.
p. 78	Adapted from 'Health Effects of Exposure to Radon BEIR VI', © 1999. Reprinted with permission by the National Academy of Science, courtesy of the National Academies Press, Washington D.C.
p. 139	Copyright Nature
p. 168	Reprinted Fig 1 – Map C only with permission from Rogers DJ and Randolph SE, The global spread of malaria in a future, warmer world, SCIENCE 289:1763–1766. Copyright 2000 AAAS.
p. 75	Roman E, Doyle, P, Maconochie N et al, 'Cancer in children of nuclear industry employees: report on children aged under 25 years from nuclear industry', BMJ, 1999, Vol 318, pp1443–50, with permission from the BMJ Publishing Group.
p. 116	Adapted from The World Health Report 2004. Changing History, pages 120–125, 2004, World Health Organization

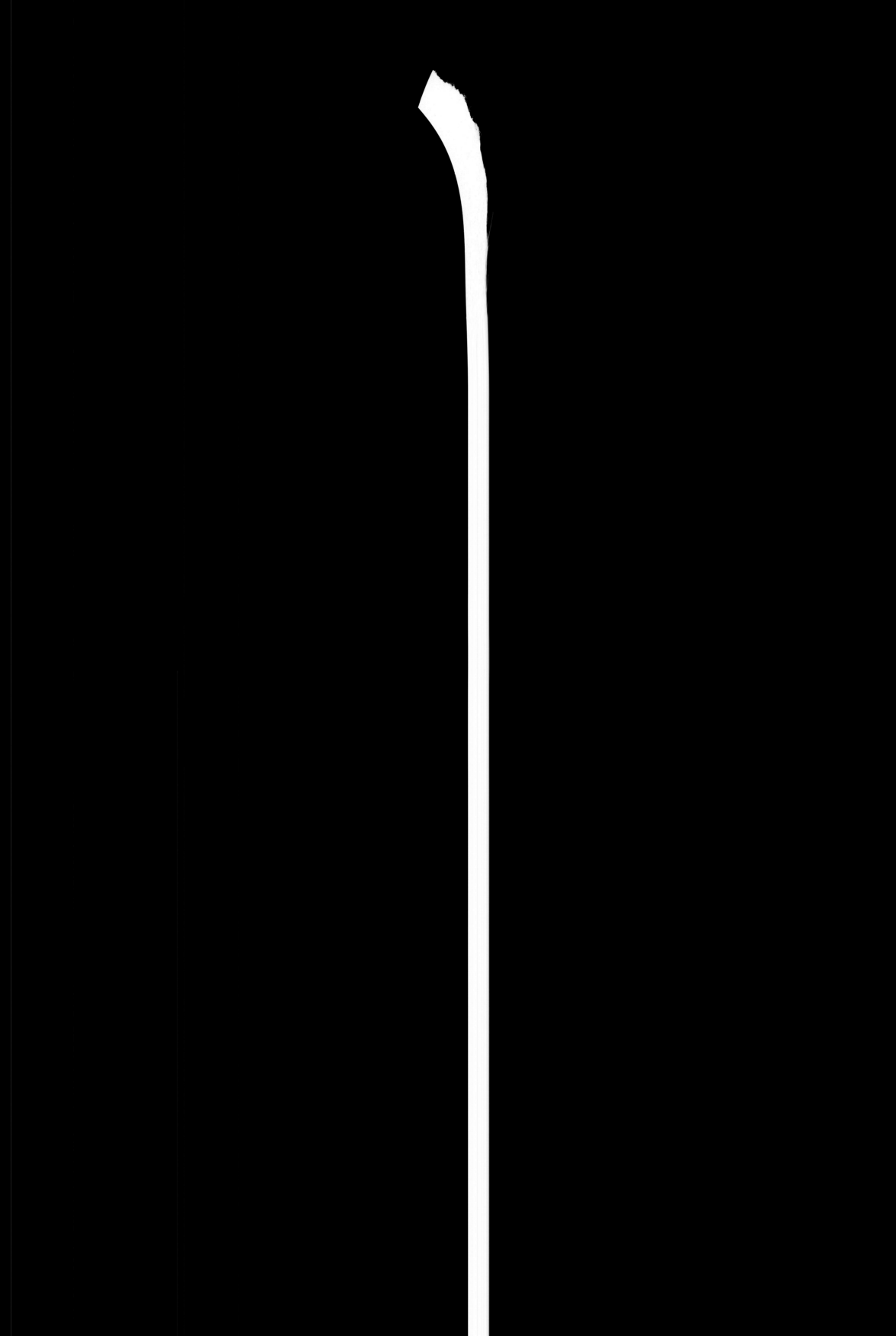

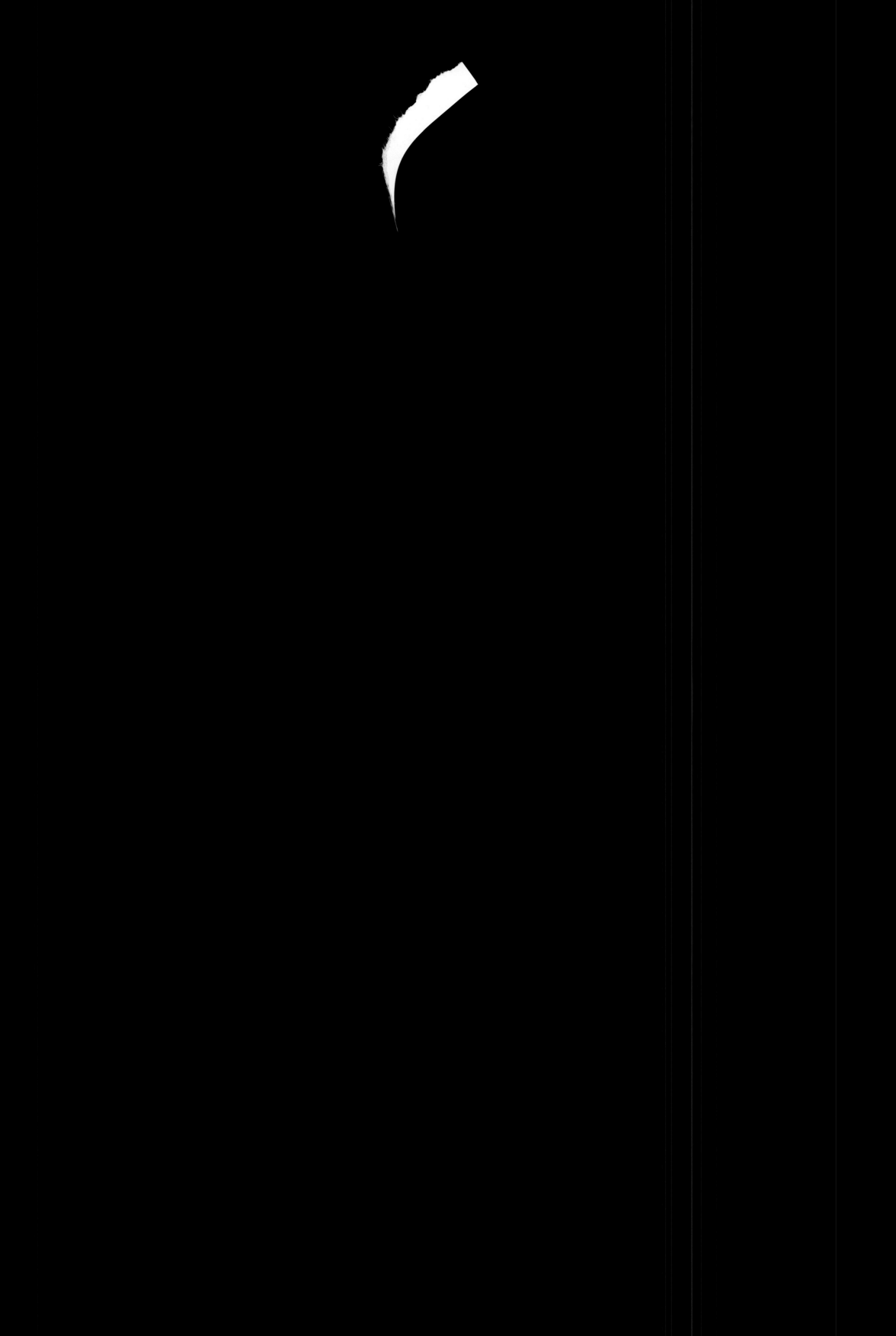

Acknowledgements

Open University Press and the London School of Hygiene and Tropical Medicine have made every effort to obtain permission from copyright holders to reproduce material in this book and to acknowledge these sources correctly. Any omissions brought to our attention will be remedied in future editions

We would like to express our grateful thanks to the following copyright holders for granting permission to reproduce material in this book.

p. 52   Anderson HR, Ponce de Leon A, et al, 'Air pollution and daily mortality in London: 1987–92', BMJ, 1996, Vol 312, pp665–669, with permission from the BMJ Publishing Group.

p. 188   Aron, Joan L., Ph.D., and Jonathan A. Patz, M.D., M.P.H., eds. Ecosystem Change and Public Health: A Global Perspective. pp309, Fig. 10.3. © 2001 The John Hopkins University Press. Reprinted with permission of The John Hopkins University Press.

p. 131   Adapted from E Cifuentes, M Gomex, U Blumenthal et al, 'Risk Factors Risk factors for giardia intestinalis infection in agricultural villages practising wastewater irrigation in Mexico', American Journal of Tropical Medicine and Hygiene, 63(3), 388–392

pp. 63–4   Reprinted from THE LANCET, Vol 360, L Clancy, P Goodman, H Sinclair and DW Dockery, 'Effect of air-pollution control on death rates in Dublin, Ireland: an intervention study', 1210–1214, Copyright (2002), with permission from Elsevier.

p. 77   Darby S, et al, 'Radon in homes and risk of lung cancer', BMJ, 2005, Vol 330, pp223–228, reproduced with permission from the BMJ Publishing Group.

p. 189   Smith GD and Ebrahim S, What can mendelian randomisation tell us about modifiable behavioural and environmental exposures?, BMJ, 330:1076–1079, reproduced with permission from the BMJ Publishing Group.

p. 59   Dockery, DW and others, 'An Association between Air Pollution and Mortality in Six U.S. Cities', 329(24), 1753–1759. Copyright © 1993 Massachusetts Medical Society. All rights reserved.

p. 102   Reprinted from THE LANCET, vol 352, H Dolk, M Vrijheid, B Armstrong et al, 'Risk of congenital anomalies near hazardous-waste landfill sites in Europe: the EUROHAZCON study,' 423–427, Copyright (1998), with permission from Elsevier.

p. 186   Reprinted from THE LANCET, Vol 360, M Ezzati, AD Lopez, A Rodgers et al, 'Selected major risk factors and global and regional burden of disease,' 1347–1360, Copyright (2002), with permission from Elsevier.

p. 154   Gouveia N, Hajat S and Armstrong B, Socioeconomic differentials in the temperature-mortality relationship in Sao Paulo, Brazil,

	International Journal of Epidemiology, 2003, 32:390–397, by permission of Oxford University Press and International Epidemiological Association.
p. 155	Hajat S, Armstrong B, Gouveia N and Wilkinson P, 'Comparison of mortality displacement of heat-related deaths in Delhi, San Paulo and London,' 2004. Epidemiology, 15(4): S94.
p. 30	Reprinted from THE LANCET, vol 360, G Hoek, B Brunekreef, S Goldbohm et al, 'Association between mortality and indicators of traffic-related air pollution in the Netherlands', 1203–1209, Copyright (2002), with permission from Elsevier.
p. 158	Kovats RS, Hajat S and Wilkinson P, 'Contrasting patterns of mortality and hospital admissions during hot weather and heat waves in Greater London, UK,' 2004, Occupational Environmental Medicine. 61: 893–898. Reproduced with permission from the BMJ Publishing Group.
p. 169	© Crown Copyright 2005 Published by the Met Office, UK
p. 97	Reprinted from Moore KL, 'The Developing Human: Clinically Orientated Embryology', 6th Edition © 1998 Elsevier Inc, with permission from Elsevier.
p. 78	Adapted from 'Health Effects of Exposure to Radon BEIR VI', © 1999. Reprinted with permission by the National Academy of Science, courtesy of the National Academies Press, Washington D.C.
p. 139	Copyright Nature
p. 168	Reprinted Fig 1 – Map C only with permission from Rogers DJ and Randolph SE, The global spread of malaria in a future, warmer world, SCIENCE 289:1763–1766. Copyright 2000 AAAS.
p. 75	Roman E, Doyle, P, Maconochie N et al, 'Cancer in children of nuclear industry employees: report on children aged under 25 years from nuclear industry', BMJ, 1999, Vol 318, pp1443–50, with permission from the BMJ Publishing Group.
p. 116	Adapted from The World Health Report 2004. Changing History, pages 120–125, 2004, World Health Organization

Overview of the book

Introduction

This book covers the principles of environmental epidemiology, drawing on examples of environmental concerns that have impacts from the local to the global. By the end of the book, the reader should be able to:

1 Describe the main methodological issues in environmental epidemiology, specifically those relating to the investigation of the heath effects of pollution of air, water and land, and the health effects of ionizing and non-ionizing radiation.
2 Assess and critically interpret scientific data relating to potential environmental hazards to health.
3 Plan, conduct and interpret the initial investigation into a putative disease cluster.
4 Describe the principles of geographical and time-series studies for the investigation of the health effects of environmental exposures, and the specific value of geographical information systems as an investigative tool.
5 Outline the evidence about global climate change and the methods for assessing its potential health impacts.

The topic area is large and this book cannot be comprehensive. The intention is rather to concentrate on methods and principles which may be applied to any environmental health hazard.

Structure of the book

The book has 15 chapters, divided into six topic sections. Each chapter, as appropriate, includes:

- an overview;
- a list of learning objectives;
- a list of key terms;
- a range of activities;
- feedback on the activities;
- a summary.

Although examples and case studies come from low-, middle- and high-income countries, the main emphasis is on high-income countries. However, the methods of investigation are applicable to most settings.

Throughout the text, we often pose as 'activities' some questions for you to reflect on. You should pause at these and write some notes of your responses before reading on to the 'feedback'. It is not expected, however, that you seek additional information to answer these, or write formal answers.

When you have thought through and noted your responses, read the feedback section. Do not be disheartened if this mentions more things than you have thought of. The feedback sections are not intended to give 'answers' that we expect you to have worked out for yourself. Rather they use the questions and your reflection on them to advance your understanding and knowledge. The following description will give you an idea of what you will be reading.

Clusters

The first three chapters look at a common issue in environmental epidemiology, namely disease clusters. Chapter 1 describes a typical example in which an apparent high risk of cancer around an industrial site is reported by a journalist. You will be asked to consider the issues raised by such a report and the importance of addressing public concern. You will also consider how you might proceed with an investigation. In Chapter 2 you will look specifically at the application of modern geographical methods for such investigation. Relevant methods of statistical analysis will be considered in Chapter 3, where you will also be introduced to the wider debate about the scientific and public health value of cluster investigations.

Air pollution

Chapters 4 and 5 consider the health effects of outdoor air pollution and the basis of the epidemiological evidence relating to those effects. You will be introduced to time-series studies, which are often applied in air pollution and health research, though their interpretation can be complex. Time-series studies provide evidence relating to short-term impacts of air pollution. In Chapter 5, you will consider the comparative advantages and disadvantages of evidence based on comparing different populations exposed to different levels of ambient pollution and meet the concept of the semi-ecological study.

Radiation and hazardous waste

Section 3 covers ionizing and non-ionizing radiation and hazardous waste. The health effects of high-dose ionizing radiation are well understood, and the current epidemiological debates centre on the effects of low-dose exposure, including cancer, genetic damage and teratogenicity. Whether typical exposure to non-ionizing radiation has health effects remains controversial, in part because the epidemiological studies present particular challenges. These will be considered in Chapter 7, while in Chapter 8 discussion turns to health effects of hazardous waste sites. The example described is that of hazardous waste landfill sites and the putative association with congenital anomalies, which is used to explore the particular features of congenital anomaly epidemiology.

Water and health

Chapters 9 and 10 consider two aspects of water and health, namely the issues relating to lack of access to clean water and, secondly, the health risks associated with the use of wastewater in agriculture. You will consider the health implications relating to the water shortage, which arise from the impact of industrialization, population growth, climate change and the global scale of the health burdens arising from inadequate access to clean water and sanitation. Chapter 10 focuses on the methodological approaches to investigating the health effects of wastewater use.

Climate change

Section 5 considers the debates about climate change and health. Climate change is the most prominent example of global-scale environmental change and it has potential impacts on ecosystems and human health. The background evidence is presented in Chapter 11, before specific examples are considered in Chapters 12 (extreme weather events) and 13 (vector-borne disease). The study of the health impacts of climate change is fundamentally different from that of studying local environmental exposures, not just because of the global scale, but also because the effects are deferred and hence the evidence is indirect and entails assumptions about the future.

Epidemiological evidence

The final section summarizes the principal issues of interpreting epidemiological evidence on the environment and health, drawing on the concepts met in earlier chapters. Finally, Chapter 15 outlines the emerging issues in environmental epidemiology and considers some of the possible future directions of research in the field.

Acknowledgements

The editor acknowledges the important contributions made by colleagues who developed the original lectures and teaching material at the LSHTM on which some of the contents are based.

Chapter 10 is based, in part, on lecture notes previously prepared by Ursula Blumenthal whose assistance is gratefully acknowledged. The editor also acknowledges the contribution of Dr Tanja Pless-Mulloli, University of Newcastle, for reviewing the text and Deirdre Byrne (series manager) for help and support in preparing this book.

SECTION I

Clusters

Investigation of a putative disease cluster

Overview

Sources of environmental pollution are often geographically localized, and so too therefore their associated health risks. The discovery of an apparent cluster of disease, for example in residents of a neighbourhood, may cause concern about an underlying environmental hazard. Such clusters frequently give rise to calls for further investigation from the public. But the subsequent enquiry presents a number of difficulties for epidemiologists and public health professionals. In this and the following two chapters you will look at the circumstances of cluster reports, the methods of their investigation and the interpretation of the resulting scientific evidence. The chapter begins by considering the concerns raised by a cluster report and the initial assessment of its public health significance.

You will first consider a case study of a cluster reported in a television documentary. You will think how you would proceed if faced with this issue as a public health specialist.

The guidelines on cluster investigations produced by the Centers for Disease Control, and cited at the end of the chapter (Centers for Disease Control 1990), are worth reading after you have worked through this and the following two chapters.

Learning objectives

By the end of this chapter you should be able to:

- **describe the immediate consequences of a report of a disease cluster**
- **propose methods for the preliminary assessment of its public health importance**

Key terms

Disease cluster An unusual aggregation of health events that are grouped in space and time.

Post hoc hypothesis Formulation of hypothesis after making the observation.

Disease clusters

A 'cluster' may be defined as 'an unusual aggregation . . . of health events that are grouped together in time and space' (Centers for Disease Control 1990). In the

Table 1.1 Some example cluster investigations that have led to advances in scientific understanding

Observed cluster/health effect	Causative agent
Bladder cancer	Azo dyes
Angiosarcoma	Vinyl chloride (Waxweiler et al. 1976)
Epidemic atypical pneumonia	Legionella (Fraser et al. 1977)
Acute exacerbation of asthma	Soya dust in Barcelona (Anto et al. 1989)

history of public health, the investigation of disease clusters has provided evidence about a range of hitherto unsuspected health risks. Perhaps the most famous example is John Snow's observation in the 1850s of the clustering of cholera cases in Golden Square in central London and his subsequent identification of a water pump in Broad Street as the common source of infection (Snow 1855). Some examples of other influential cluster studies are listed in Table 1.1.

Episodes of food poisoning and outbreaks of other forms of food- and water-borne disease can be considered disease clusters. But in general their investigation does not focus on geography so much as on shared food sources and personal contacts, and they are distinct in a number of ways from non-communicable disease clusters with a possible environmental cause. This chapter discusses the specific circumstances of non-communicable disease clusters and their linkage to environmental exposures.

We begin by considering the case of a putative cluster that was first brought to public attention through a television documentary. We will elaborate the stages of investigation of this cluster in the three chapters of this section. It is based on a real-life example, but we have amended various parts of the story and evidence to illustrate the principles of cluster investigation. A paper of the original study has been published (Wilkinson et al. 1997).

Case study: cancer risk around a pesticide factory

Concerns of a possible cancer cluster were first raised when an investigative journalist found an apparently high number of cancers among workers and residents living in the vicinity of a pesticide factory in Britain. Attention was focused on two roads bordering the plant (highlighted in bold in Figure 1.1). A television documentary was produced which contained a series of interviews with cancer victims and their relatives, and with the managing director of the factory. Although there was some uncertainty over the exact number of cancer cases, the programme appeared to report at least eight cases in the roads bordering the plant over a period of ten years or so. They included:

Brain cancer	3 cases
Lung cancer	2 cases
Malignant melanoma	2 cases
Pancreatic cancer	1 case

Site location

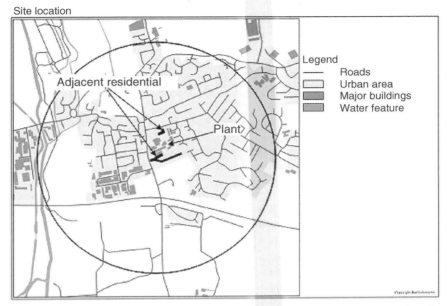

Legend
—— Roads
☐ Urban area
■ Major buildings
■ Water feature

Adjacent residential

Plant

Figure 1.1 Local area map of the factory and neighbouring streets. The circle has a radius of one kilometre and is centred on the plant

Activity 1.1

Read the edited transcript below and then make bullet point notes of your response to the following questions as if you were the public health specialist responsible for the population in which the factory is located:

1 What is your immediate assessment of the seriousness of the health hazard around the pesticide factory and the impact of the documentary?

2 What features of the cluster and its reporting are most important in your assessment of what action to take?

3 How sure are you of being able to establish or refute a link between the plant and illness in local residents by further enquiry?

Edited transcript from television documentary

'In the last ten years in this street cancer has killed at numbers 5, 25, 33, 37 – twice – 43, 44, 45 and 51. The street runs alongside one of Britain's major pesticide factories. Is there a connection? Tonight we reveal evidence of a cancer cluster amongst the factory workers and people living nearby.'

The documentary then describes the cases of two employees who had developed and died from cancer.

'If they were one-offs, there would be nothing unusual about [these] deaths, but they're not. We made a detailed study of cancer deaths since 1982 amongst workers whose jobs brought them into direct contact with the formulation of pesticides. We traced a total of seven men who have died from various forms of cancer. There may be more. But even this number is three times higher than would be expected. We took our preliminary findings to a leading occupational epidemiologist: 'We have looked at the numbers of workers, the age distribution of those workers, and so from that one can estimate more or less how many you'd expect amongst the workers in this factory, and in fact you'd expect to find one or two cancers and we found seven amongst the males, that's a significant . . . a statistically significant excess.'

↻ Feedback

1 At face value, the documentary appears to provide evidence of a serious cancer risk, which is bound to create concerns among local residents and the workforce even without further substantiation. Most viewers are likely to be persuaded by it, and some may view the issue as one of industrialists against a workforce who are suffering health consequences because of insufficient investment in industrial hygiene. The fact that the putative cluster was reported in a television documentary raises the stakes and has a number of immediate consequences, irrespective of the underlying truth:

- property prices in the area may well have fallen
- there will be immediate concerns among local residents as well as the workforce
- legal action by the families of cancer victims or other affected people is a possibility
- question marks may be raised in relation to the operation of the company which is probably one of the largest local employers

2 The cancer cluster itself is not well defined in time and place, it covers a number of different pathological sites, including bowel, brain, lung and skin, and the candidate causative agent(s) and route(s) of exposure are unclear. These factors make it more difficult to advance a specific hypothesis, and less likely that there is a genuine cluster caused by an occupational or environmental hazard relating to the plant.

3 At face value, one would guess that it should not be difficult to assemble firm evidence. However, experience tells us that investigation of disease clusters such as this rarely, if ever, produces clear evidence of a cause and effect relationship.

✏ Activity 1.2

Most of the population living within a kilometre of the plant is contained within two census areas (known as 'wards'), whose populations can be looked up from routinely available tables. The wards are HNBF, with a population of 3039 men and 3099 women, and HNBG containing a resident population of 2483 men and 2571 women. Tabulations of cancer registrations are also available, and some selected (all ages) registration rates are given below (Table 1.2).

Using these data and the little evidence you have heard from the documentary, consider how many cases of cancer were reported and how many you might expect in the local area and among workers. Do you think that the occurrence of cancers demands, as the programme makers suggest, a full and systematic enquiry?

What more information would you like to better inform your answer?

Table 1.2 Cancer registration rates per 100,000 population, England and Wales, all ages

	Male	Female
All malignant neoplasms	490.5	464.9
Stomach	28.2	16.7
Colon	31.1	36.2
Rectum, recto-sigmoid junction and anus	23.6	17.6
Pancreas	12.5	12.3
Trachea, bronchus and lung	108.0	45.6
Breast	0.8	103.4
Brain (malignant)	7.4	1.5
Brain and nervous system (benign)	1.4	2.4

 Feedback

1 Cluster among workers? You have no information on the number of workers, so it is not possible to calculate expected cases yourself. You have little evidence to go on except for the testimony of the epidemiologist that there were one or two cases of cancer expected among male workers and seven observed – an apparent excess.

You would however like to have much more information before deciding whether this amounts to *prima facie* evidence of a cluster. Some of the issues to consider include: *how the cases were ascertained and what their confirmed diagnoses are*. The reported cancers include at least four separate pathological sites which indicates little specificity, and argues against a specific cause and effect. *What chemicals are handled at the plant and how and when people might have been exposed to them*. Relevant here is date of diagnosis in relation to date of starting work at the plant. The minimum latency for solid tumours, including lung and brain, is generally considered to be five or ten years; it is less for other malignancies, such as leukaemia. So more recent exposures are very unlikely to give rise to such tumours. Although the factory doubtless handles a range of chemicals it would be useful to know which if any have been identified as animal and/or human carcinogens. *How the calculation was made of the number of expected cases*. You have only the evidence of the documentary, and you don't know if the editors may have been selective with their evidence to construct the best story.

2 Cluster in nearby residents. Similar considerations apply in relation to assessing the significance of cases among local residents. However, because we have some population data, we can do a rough calculation of the expected numbers of cancers to indicate the likelihood that an important excess has occurred. If we take the two census wards together, and ignoring age (defensible for a general residential population), we can calculate the expected numbers of cases as shown in Table 1.3.

Although it is a little unclear, the documentary suggested probably nine or more cancers *among residents* in nearby roads, of which three were brain tumours. Clearly the expected number of cancers in total for the two wards was far in excess of the numbers observed in just the two roads next to the plant, but then the wards include many more roads than just those two. The analysis by ward is probably too crude to be useful, even though (in Britain) these are the smallest areas for which cancer statistics can usually be obtained.

Table 1.3 Approximate calculation of expected numbers of cancer cases in two census areas (wards) around the factory

		Rate/100,000		Expected no. over 10 years	
	Population at risk	All cancers	Malig. brain	All cancers	Malig. brain
	a	b	c	(a*b*10)/100,000	(a*c*10)/100,000
Men	3039 + 2483 = 5522	490.5	7.4	270.9	4.1
Women	3099 + 2571 = 5670	464.9	1.5	263.6	0.9
Total				534.5	5

However, it is striking that the total number of cancers is much less in relation to the expected total than the observed number of brain cancers is in relation to the total expected brain cancers. This can be somewhat formalized with an *ad hoc* proportional registration analysis. The proportion of cancers which are of the brain (3/9 = 33.3 per cent) is much higher than the proportion of cancers in England and Wales from Table 1.2 (7.4/490.5 = 1.5 per cent in men, 5.4/464.9 = 1.2 per cent in women, about 1.3 per cent overall), a proportional registration ratio of about 33.3/1.3 = 26. Following this rather *ad hoc* logic, on this proportional basis, we expect about 0.013 × 9 = 0.12 brain cancers. We cannot draw clear conclusions about the number of cancers overall, but the proportion of these that are brain cancers does appear to be high to an extent that would make it a very unusual event.

Another way of approaching this could be to do very rough calculation of the occurrence of cancer in the two roads adjacent to the factory. The highest house number with a reported cancer was no. 51 (in the longer road), so we might guess that the roads together contain around 100 dwellings. At usual UK occupancy rates, these dwellings might therefore contain about 250 residents. The expected number of brain cancers = ((7.4 + 1.5)/2)/100 000 (rate in general population, both sexes combined)) * 250 (people) * 10 (years) = 0.11 case. So, in these roads, the observation of three cases provides indication that brain cancer occurrence is much higher than expected.

Other considerations: *post hoc* hypotheses

In current epidemiological usage, *post hoc* refers to the formulation of a hypothesis after the event (on the basis of observed data). In this example, having seen an 'excess' of cancer cases – the reported 'cluster' – we are formulating a hypothesis that there is an excess of cancer cases caused by the plant. This has an element of circularity to it, and it shouldn't be surprising that we find evidence of an excess if statistical tests are applied.

In fact, this is equivalent to multiple testing. You choose to test this area precisely *because* there appears to be a lot of disease there. But you are really being highly selective in testing only this particular area. There is in fact almost an infinite array of different sets of cases that we could test based on varying boundaries of time, space, diagnosis etc. We selectively test the one combination that appears unusual. But it may be unusual for no other reason than a chance occurrence in the random variation of disease. Other inconspicuous aggregations of cases simply aren't tested.

This somewhat philosophical issue is fundamental to the interpretation of statistical inference and will be further discussed in Chapter 3. Its importance lies in the fact that if we generate our hypothesis after the event (*post hoc*) it is impossible to make a proper interpretation of tests of statistical inference. On the other hand, if the hypothesis was generated before seeing the apparent cluster, then inference is more secure. It all depends on the circumstances in which the cluster came to light.

 Activity 1.3

What other action, if any, should be taken to safeguard the health of the residents now?

 Feedback

Judgements need to be made about how great is the potential threat to human health, if any, and whether workers and residents continue to be exposed to that threat. If so, decisions need to be taken about:

- removing exposures by changing hygiene practices, closing parts of the works (or even the whole of it), decontaminating land etc.
- informing workers and residents of what precautionary actions and further investigations have been set in train
- screening health checks for cancer to workers and local residents
- environmental sampling to test for contamination
- informing other local and national authorities so they may also act as necessary
- engaging the community in further discussion of the potential health risks, protective action(s) and further investigation

However, apart from inspecting the plant to ensure it meets required standards of hygiene, the evidence probably does not demand other action at this stage. The benefits of any precautionary action have to be balanced against the potential harm to the local economy, people's livelihoods etc.

Centers for Disease Control (CDC) guidelines for investigating clusters

Deciding how to proceed with cluster investigations can be difficult, and a balance approach is required. The CDC guidelines (1990), which are broadly followed by most official agencies with responsibility for cluster investigations (e.g. the Agency for Toxic Substances and Disease Registry), suggest a number of stages to the cluster enquiry. The initial phases may include:

- establishing a case definition (needed for two reasons: for epidemiologic surveillance studies relating to the prevalence of the disease, and also for diagnostic purposes using applicable diagnostic features, causes and patho-physiology);
- confirming the reported cases;
- defining the population denominator/expected number;
- reviewing the published scientific literature;
- carrying out exposure assessment;
- generating and testing biologically plausible hypotheses;
- communicating the results.

 ## Summary of CDC guidelines for investigating disease clusters

Clusters of health events, such as chronic diseases, injuries and birth defects, are often reported to health agencies. In many instances, the health agency will not be able to demonstrate an excess of the condition in question or establish an etiologic linkage to an exposure. Nevertheless, a systematic, integrated approach is needed for responding to reports of clusters. In addition to having epidemiologic and statistical expertise, health agencies should recognize the social dimensions of a cluster and should develop an approach for investigating clusters that best maintains critical community relationships and that does not excessively deplete resources.

Health agencies should understand the potential legal ramifications of reported clusters, how risks are perceived by the community, and the influence of the media on that perception. Organizationally, each agency should have an internal management system to assure prompt attention to reports of clusters. Such a system requires the establishment of a locus of responsibility and control within the agency and a process for involving concerned groups and citizens, such as an officially constituted advisory committee. Written operating procedures and dedicated resources may be of particular value. Although a systematic approach is vital, health agencies should be flexible in their method of analysis and tests of statistical significance. The recommended approach is a four-stage process: initial response, assessment, major feasibility study and etiologic investigation.

Each step provides opportunities for collecting data and making decisions. Although this approach may not always be followed sequentially, it provides a systematic plan with points at which the decision may be made to terminate or continue the investigation.

With respect to further epidemiological investigation, one might:

- assess the potential of exposures emanating from the plant to give rise to cancer, in particular brain cancer;
- find expected numbers for residents living 'close to' the plant, using a better definition than simply living in these two wards;
- ascertain all cancers in people living close to the plant, and perhaps in a nearby control area;
- do an epidemiological study of cancer in areas close to similar plant(s);
- look for things which distinguish the cases from others in the cluster area apart from residence there;
- investigate whether there is a dose-response relationship with exposure within the cluster area.

In the next chapter you will look at ways in which a small area study might be done of cancer incidence and mortality in the locality using modern methods of geographical analysis.

Summary

In this chapter you have looked at the questions raised about the observation of an apparent disease cluster, taking as a case study cancer incidence in workers and residents in the vicinity of a pesticide factory in the UK. Immediately news of such an observation is made public, concerns are bound to be raised among workers and the local community, and a range of consequences follow irrespective of whether a 'true cluster' is later substantiated. The immediate assessment of the cluster is often difficult from routinely available sources of data, so decisions have to be made about public health protection and further investigation on the basis of incomplete evidence. The imperative for further investigation may be demanded as much by the need to address public concerns as by the scientific case. The form of further investigation and action, which may bring in the participation of a range of experts and representative bodies, will be considered in the next two chapters.

References

Anto J et al. (1989). Community outbreaks of asthma associated with inhalation of soybean dust. New England Journal of Medicine 320(17): 1097–102.

Centers for Disease Control (1990). Guidelines for investigating clusters of health events. MMWR 39(RR–11): 1–23.

Fraser D, Tsai T et al. (1977). Legionnaires' disease: description of an epidemic of pneumonia. New England Journal of Medicine 297: 1189–97.

Snow J (1855). On the Mode of Communication of Cholera. London, John Churchill.

Waxweiler R, Stringer W et al. (1976). Neoplastic risk among workers exposed to vinyl chloride. Annals of the New York Academy of Science 271: 40–8.

Wilkinson P, Thakrar B et al. (1997). Cancer incidence and mortality around the Pan Britannica Industries pesticide factory, Waltham Abbey. Occupational and Environmental Medicine 54(2): 101–7.

Useful websites

Centers for Disease Control, Atlanta, GA: www.phppo.cdc.gov/cdc

Eurocat (Congenital Anomalies and Public Health). www.eurocat.ulster.ac.uk/

Agency for Toxic Substances and Disease Registry (ATSDR): www.atsdr.cdc.gov

National Institute for Occupational Safety and Health (NIOSH): www.cdc.gov/niosh

US Environmental Protection Agency (EPA): ://www.epa.gov

2 | Geographical analysis of an industrial hazard

Overview

This chapter considers the use of modern geographical methods, specifically geographical information systems (GISs), for analysing health data in relation to sources of environmental pollution. It is based on data which simulate a cluster around an industrial site. It illustrates methods relevant to the further investigation of the sort of cluster report considered in Chapter 1 (assuming further investigation is warranted). These methods are, however, equally appropriate to scientific studies addressing hypothesized risks associated with putative industrial or other environmental hazards. Although GIS-based methods have several advantages, limitations of data and design are important to bear in mind in the interpretation of results. You will consider issues of statistical analysis and interpretation in Chapter 3.

Learning objectives

By the end of this chapter you should be able to:

- describe the principles of **GIS** methods for investigating environment and health issues
- describe the strengths and weaknesses of these methods

Key terms

Geographic Information System (GIS) An information system used to store, view, and analyse geographical information.

Raster A form of spatial data representation in which the data are stored as a matrix of cells or pixels.

Vector (in mathematics and physics) A quantity having both direction and magnitude which determines the position of one point in space relative to another.

Geographical information systems

We begin with a short description of a GIS, which may be defined as a computer system capable of holding and using data on geographical objects or, more generally, as a combination of hardware, software and personnel, capable of storing, editing, analysing and displaying geographically referenced data.

At the heart of a GIS is software specifically designed to make it easy to analyse spatial relationships between geographically-coded features. It can therefore be used to produce maps, to calculate distances, to define adjacency of features, or to carry out more complex analysis, such as the computation of the local density of a particular feature.

Within a GIS database, any geographical object has associated with it two types of information:

- spatial (i.e. its location using some coordinate system);
- attribute (i.e. the characteristics of the object or what it represents) (Figure 2.1).

It is the combination of information that makes GIS a powerful tool, as it allows datasets ('map features') to be superimposed on top of each other and for distances and spatial relationships to be computed between them. Geographical features are split into layers, each of which contains only one type of feature (e.g. soil type, land use, roads, rivers, administrative boundaries). Features can be represented within each layer as points, lines, polygons (areas with an identifiable boundary) or images (Figure 2.2). GIS can store data either in raster format (where data are represented by a regular grid, typically used for handling satellite data) or as vectors (quantities with both magnitude and direction which determine position in space).

Location data are recorded with reference to a coordinate system, which is defined by an origin (which places points relative to the earth's surface) and by its units of measurement. Most coordinate systems assume a rectangular grid, but this is an oversimplification for defining position on the nearly spherical earth's surface. Taking points from a spherical object and transforming the coordinates onto a grid produces errors; as you move away from the origin the larger the errors become. Most coordinate systems are therefore considered to be 'local', and it is important to choose the correct coordinate system for the region of interest. (Latitude and longitude, however, are units of position on a spherical surface and thus do not suffer from this problem.)

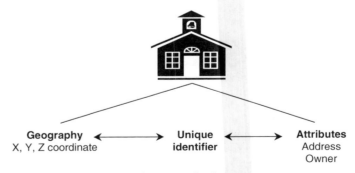

Geography ←→ **Unique** ←→ **Attributes**
X, Y, Z coordinate   **identifier**   Address
                                                          Owner

Figure 2.1 Information associated with a geographical object

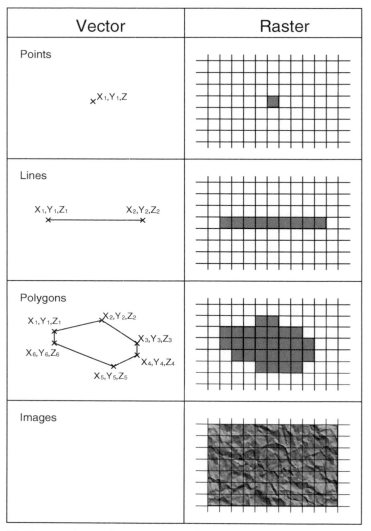

Figure 2.2 Vector and raster feature representation

Knowing the origin and units for a coordinate allows it to be related to the earth's surface but there is still the issue of how to display a spherical surface as a flat map. The form of this display is known as the map projection – which is often considered to be the third element of a coordinate system. Each map projection has its own properties and all produce distortions of distance, direction, scale and area. Some projections minimize distortions in some of these properties at the expense of larger errors in others. A form of projection used on many world maps is the Transverse Mercator, in which the earth's surface is projected onto a cylinder. The coordinate system for the UK National Grid is defined as follows:

- projection: universal Transverse Mercator;
- origin: south-west England;
- units: metres.

Data

The gain from using GIS depends on the user's abilities, the functionality of the package and, most importantly, the available data, which will vary by country and even individual user. GIS data may be obtained from several sources:

- *Archives of digital data.* It is always worth asking about the availability of such data as they can save much time and effort. Research institutions, local agencies, government bodies and commercial companies are all possible sources. Sometimes the datasets cover large areas or are very detailed and so would be too large to generate yourself, i.e. census boundaries for England.
- *Digitizing and scanning.* Where no digital dataset can be found, or the available data is just too expensive, new datasets can be created by digitizing or scanning. These processes involve defining the objects of interest on a paper map, aerial photograph or satellite image and then manually 'tracing' them to create the digital version. Because this process has to be carried out manually, it is time-consuming and so usually only used for small sets of data, i.e. road network or land use in one district.
- *Global positioning systems.* Global positioning systems (GPSs) use a network of satellites to locate any point on the earth's surface to an accuracy of 1 to 100 metres. Collection of geographical locations using a GPS is particularly useful in remote areas where other data sources may be sparse, especially if locations are being visited anyway to collect other types of data, i.e. a village location in Africa.

The advantage of GIS is that it allows large sets of geo-referenced data to be spatially analysed with comparative sophistication and ease using either 'off the shelf' data or data which the investigator has generated, or both. It is particularly valuable for analysing health risks in relation to environmental hazards where routine sources are available (e.g. post-coded mortality data which can be spatially related to 'pollution maps').

Case study

You will now consider the example of a putative cancer cluster around the industrial site.

 Activity 2.1

Look at the GIS-generated map shown in Figure 2.3 which shows the site and distribution of cancer cases around it. The cases and their locations, based on residential address, were supplied by the cancer registry for this area.

1   What can you infer about the possible cancer risk associated with the site?

2 What do you think is the main determinant of the distribution of cases?

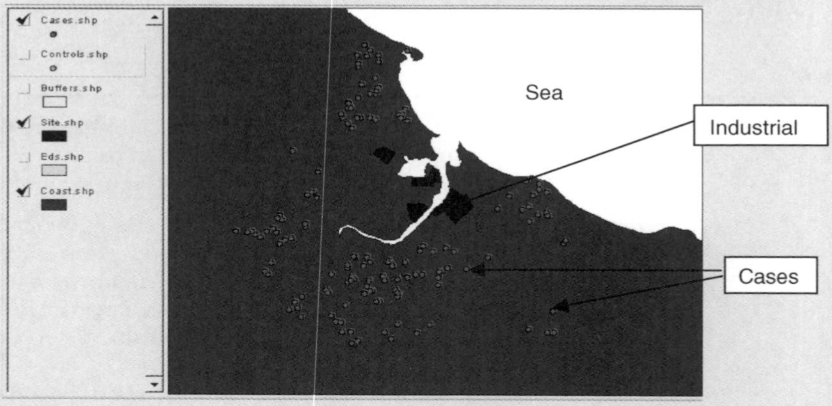

Figure 2.3 GIS-generated map showing the coastline, location of the industrial site (solid shading) and cancer cases (dots) in the local area

Feedback

1 Although the cases are clearly located around the industrial plant, there is very little that you can conclude about the level of hazard, if any, from the plant. The distribution of cases may or may not be influenced by emissions from the plant, but to judge cancer risk the variation in cases needs to be related to the population at risk.

2 Regardless of the specific risk factors, the main determinant of the distribution of cases is, of course, population density: cases can only occur where people live. In this example, the fact that cases appear clustered around the plant simply reflects the fact that the industrial site is located on an estuary with built-up areas around it.

Activity 2.2

Given the need to know about the underlying population: 1. What approaches might you use to examine local variation in risk, 2. How would you go about obtaining the relevant data?

Feedback

You might consider using two broad methods for looking at local variation in risk:

1 Linkage of these cases to areas for which you have population data, e.g. from the census or other sources (Figure 2.4).

2 Obtaining a set of controls to reflect the distribution of the population (Figure 2.5).

Each method has its own advantages and disadvantages.

Area population data provide the basis for computing absolute rates of cancer incidence or mortality, which could, in theory, be compared with rates in other areas. Population data are often available from published sources.

However, in many countries they are available only for quite large areas and the areas are typically defined by administrative boundaries which may have little relevance to the environmental hazards in question. A further disadvantage is that there is often little information about the population other than its size by age and sex, and this can be problematic in the common situation of being concerned about potential confounding.

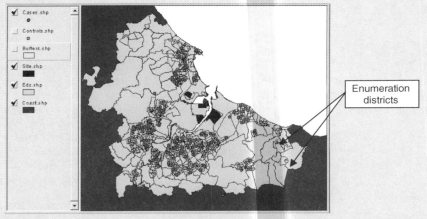

Figure 2.4 Maps of cases (light grey dots) superimposed on the boundaries of areas for which population data are available (these census areas are known as enumeration districts)

Figure 2.5 Map of cases and controls in relation to the industrial site

The advantages of selecting a set of controls include the fact that they can provide point data for comparison with the location of cases and can be analysed more flexibly. It may also be possible to gather more information about the characteristics of cases and controls. But, such data are less readily available, and may require careful selection from a population register followed by survey.

Activity 2.3

1 What criteria would you use for selecting the control population in this type of study?

2 How would you measure exposure?

Feedback

1 As in any case-control design, the cases and controls are selected on the basis of disease status alone. Remember that the purpose of the controls is to represent the exposure in the population from which the cases have come. It is tempting to consider drawing controls from the same street as the cases (neighbours) but this would not be appropriate if the location of residence is used to categorize exposure: it is the difference in exposure (and thus of location) between cases and controls that we wish to test. You might legitimately match on age or sex, for example, but not on location. The only geographical restrictions with the current data are that both cases and controls have been constrained to come from within ten kilometres of the industrial plant.

2 With regard to exposure, ideally one would want to use a direct measure of exposure, based on personal assessment. A second best would be to obtain a measure of pollution concentration where the individuals live. Clearly, if we have specific data about pollutants and their dispersal patterns, we could generate contours of pollution concentrations and superimpose them on the map of cases and controls within the GIS. Individuals could then be classified according to the value of the pollution concentration at their place of residence. In practice, there is often no good data about how pollution concentrations vary around the site, nor even, in many cases, about which pollutants are of specific concern. In these circumstances, researchers often use distance as a simple proxy. This is based on the assumption that living close to the site carries a higher level of exposure regardless of the pollutant or route of exposure. The computation of the distance to the nearest point of an extended pollution source such as the industrial area can be accomplished with comparative ease within a GIS, as illustrated by the distance bands in Figure 2.6.

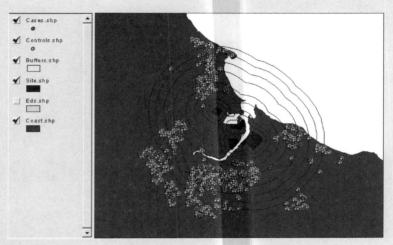

Figure 2.6 Map of cases and controls showing distance bands around the site

 Activity 2.4

So far, then, you have a basic set of case-control data with distance from the site as the surrogate measure of exposure. What other information would you like to have before you begin your analysis?

 Feedback

The obvious deficiency at present is any data about confounding factors. We are likely to have data on the age and sex of the cases and controls, and can therefore adjust for them. But we should also be concerned about other potential risk factors, including, for example, socioeconomic status. If special surveys are carried out, data could be collected about risk factors at individual level. However, even if no surveys are used, socioeconomic data may be available for areas, such as the enumeration districts shown in Figure 2.4. By attaching the socioeconomic classification of the area of residence to individual case and control records, we have a simple marker of socioeconomic status that can be used in adjusted analyses (Figure 2.7).

You will see the utility of this in the next chapter which looks at statistical analysis of these data.

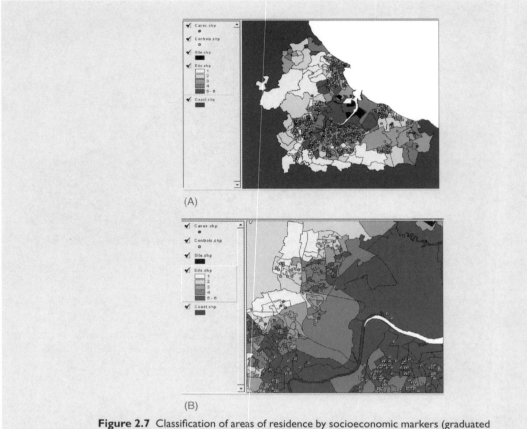

(A)

(B)

Figure 2.7 Classification of areas of residence by socioeconomic markers (graduated shading). These area markers can be linked to cases and controls within the GIS. (A) study area; (B) higher resolution view to show overlaying of datasets

Data checking

A further advantage of GIS is that it allows you to explore data interactively to understand their spatial features. This may be useful when checking for data errors.

An example is shown in Figure 2.8, which shows a plot of the number of cases at individual postcode locations. Ordinarily you would expect only one or occasionally two cases to occur at the same point location. Here counts of cases have been made by postcode, and in the UK only around 14 households share the same postcode. But the GIS shows that at one postcode there are more than three case registrations (in fact there were nine), which seems high.

By more detailed interrogation of the data it was possible to determine that this postcode is a hospital whose address was sometimes used by recording clerks whenever they had no postcode of home address for a case they were registering.

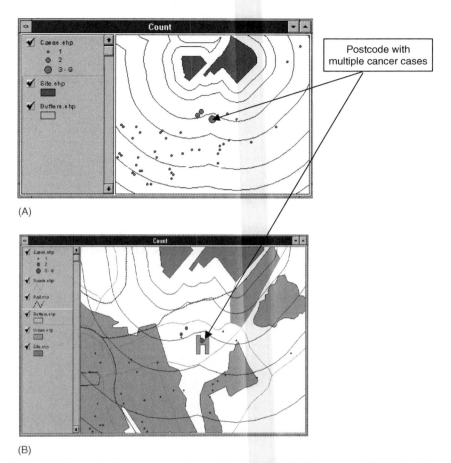

Figure 2.8 Example of interactive analysis of the data using GIS. The postcode highlighted in (A) has several cancer registrations, and overlaying other datasets shows this to relate to a hospital (here represented by an H-shaped building)

This is an important potential source of bias when analysing data at the small area level as just a few cases can substantially alter the pattern of results. Interactive analysis of the data can help identify such problems.

As a result of the GIS preparation, we end up with a dataset of cases and controls classified by exposure (distance from the industrial site) and socioeconomic status. As a final stage, it would be usual to export the data from the GIS to a statistical package in readiness for formal statistical analysis. You will go through the steps of this analysis in the next chapter.

Summary

You have looked at the use of GIS to prepare data for a study of cancer risk in relation to a putative source of environmental exposure. GIS analysis is specifically designed to facilitate spatial processing, and it is particularly valuable for handling large sets of geo-referenced data obtained from routine sources.

Here you used cancer registry data and plotted the cases in relation to the site using the place of residence as the marker of location. To obtain a measure of spatial variation in cancer risk you could either relate these cases to population data available at small area level (analysis of rates) or use a case-control design. Exposure classification can be made by overlaying pollution data within the GIS, though in many cases a simple distance parameter is used as a surrogate. Markers of socio-economic status available at small area level can be used to control for confounding by socioeconomic factors in subsequent statistical analysis. The GIS can also be helpful in the interactive exploration of data.

Despite its many advantages, analyses based on GIS methods are often limited by the availability of data, particularly with regard to individual-level confounding factors and detailed measures of exposure.

3

Analysis and interpretation of a single-site cluster

Overview

You will now continue the analysis of the data you began to look at in the last chapter on geographical methods. GIS is useful to map the cancer cases and information about the populations from which they arise and to calculate measures of proximity to the industrial plant. But to investigate formally whether there is evidence for an increasing risk of disease close to the site, you need to analyse the data statistically.

You will look at this in this chapter, which goes through various steps of statistical analysis, and then concludes with brief discussion of the current debates about the investigation of clusters. It would be worth reading the reference notes given in Appendix 2.

Learning objectives

By the end of this chapter you should be able to:

- **carry out simple statistical analyses of health data in relation to a point source of environmental exposure**
- **perform tests of association between the source and disease occurrence**
- **describe some of the difficulties of interpreting the significance of such tests**

Key terms

Texas sharp shooter phenomenon A term used to refer to *post hoc* studies: the Texas sharp shooter shoots first, then draws the target where most bullets have hit. The epidemiological analogy is the selection of a cluster from the pool of all potential clusters.

Introduction to analysis

The analyses in this chapter are based on fictional data generated using the GIS methods described in Chapter 2. They are point data showing the place of residence of 150 cancer cases and 750 controls from the local area around the industrial plant.

Before beginning the formal statistical analysis of the case-control data, it was decided to look at the number of cases within two kilometres of the plant. Using cancer registration data it was possible to determine that within this distance there were 9 cancer cases and 2.88 expected from age-specific national rates (computed using indirect standardization). Could this excess be due to chance?

An indication of the role of chance in such cases can be determined by calculating a z-score using the formula $z = 2 (\sqrt{D} - \sqrt{E})$ where D is the observed number of cases and E the expected number for the chosen population and time period (see Appendix 1). D is assumed to be an observation from a Poisson distribution with mean $\mu = \theta E$ where θ is a measure of the size of excess risk. If there is no excess risk $\theta = 1$. A one-sided p-value for $\theta = 1$ is obtained from the probability of observing D or more cases in a Poisson distribution with $\mu = E$ (or approximately from $z = 2 (\sqrt{D} - \sqrt{E})$ using tables of the normal distribution). More informative than a significance test is the estimated value of θ, given by $\theta = D/E$. A 95 per cent confidence interval for this ratio is given by the formula $(\sqrt{D} \pm 1.96/2)^2/E$.

Activity 3.1

Using the formula above for the z-score ($z = 2(\sqrt{D} - \sqrt{E})$), calculate a p-value for the number of cases within two kilometres of the plant and interpret the result in each of two scenarios:

1 If the investigation was instigated because residents near a similar plant had experienced an excess of this disease.
2 If the investigation followed the observation by an alert receptionist at the oncology unit of the local hospital that several patients suffering from this cancer came from this area of the city.

Feedback

Substituting the numbers from this example, we obtain:

$$z = 2(\sqrt{9} - \sqrt{2.88}) = 2.61, p=0.01 \text{ (two-sided)}$$

The interpretation of this p-value differs depending on the scenario. In scenario (1) we conclude that the excess is very unlikely to have occurred by chance. In (2), however, the interpretation is difficult as it is unclear how many other 'non-clusters' this or another oncologist may have passed by before bringing this one to the attention of investigators. Hence the p-value does not properly reflect the play of chance. This is a typical *post hoc* analysis that you learnt about in Chapter 1.

Analysis of case-control data

The following pages take you through the steps of a more formal statistical analysis of the point data of the place of residence of the cancer cases and controls. The analysis was run using the statistical package Stata, whose commands and output are shown. The variables contained in the analysed dataset are:

x	the x coordinate of the cases and the controls
y	the y coordinate of the cases and the controls
case	1 if the subject is a case, 0 if control
netdist	distance (kilometres) from the subject's residence to the nearest part of the site ('gross' distance is distance from centre of site – more on that later)
depriv	the level of deprivation: 1 (least deprived), 4 (most deprived) of the enumeration district of the case/control
x source	x coordinate of the centre of the source
y source	y coordinate of the centre of the source statistical package.

Summary statistics of these data may be shown using Stata's *summarize* and *tabulate* commands:

```
. summarize

    Variable |     Obs        Mean    Std. Dev.        Min        Max
-------------+--------------------------------------------------------
           x |     900     1451078    5479.614     1441805    1465905
           y |     900     1521764    5664.289     1513105    1535805
        case |     900    .1666667    .3728852           0          1
     netdist |     900    5.792489    2.329267    1.218003   9.998764
    x source |     900     1452640           0     1452640    1452640
    y source |     900     1524570           0     1524570    1524570
      depriv |     900    2.498889    1.118655           1          4

. tab case

       case |      Freq.     Percent        Cum.
------------+-----------------------------------
          0 |        750       83.33       83.33
          1 |        150       16.67      100.00
------------+-----------------------------------
      Total |        900      100.00
```

These confirm that there are 900 records: 150 are cases (case = 1) and 750 controls (case = 0) – i.e. five controls per case. The mean distance from the plant is 5.79 km, the minimum distance 1.21 km and the maximum 9.99 km.

 Activity 3.2

What graphs might you now generate to explore whether the risk of cancer is increased near the plant?

 Feedback

Your aim, of course, is to compare the distribution of cases and controls in relation to the industrial site. To do this you might consider a number of options. A first approach

might be to plot in two dimensions the location of cases and controls (i.e. a map)
(Figure 3.1).

```
. graph y x, by(case) symbol([case])
```

(This means graph y against x separately for cases and controls and use display
symbols of 1 for cases and 0 for controls.)

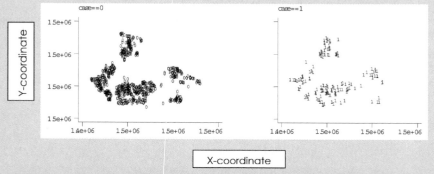

Figure 3.1 Stata commands and output to plot case and control locations

This shows no obvious difference in the distribution of cases and controls, which is
unsurprising. The main principal determinant of the spatial distribution of both is likely
to be population density – cases occur where people live!

More discriminating might be to show the distribution of distance from the site
(netdist) as box plots or histograms (Figure 3.2).

```
. graph netdist, by(case) box
```

(This means generate a box plot of netdist by case-control status.)

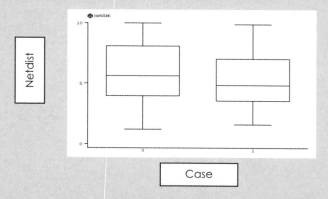

```
. graph netdist, by(case) hist bin(10) xlabel
```

(This means generate histograms of netdist separately for cases and controls.)

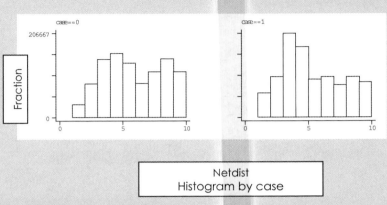

Netdist
Histogram by case

Figure 3.2 Box plots and histograms

These latter two sets of plots provide some suggestion that the cases are located slightly nearer to the industrial site than controls: the median distance is shorter for the cases (box plot) and the histograms suggest that there might be a slight preponderance of cases within the first few kilometres of the industrial plant. But you would not want to rely on informal judgement to assess whether this pattern could be due to chance.

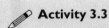 **Activity 3.3**

How might you more formally examine whether cases are indeed located nearer to the site?

Feedback

The simplest thing would be to tabulate the mean distance from the site of cases and controls, and then compare them with a two-sample t-test (or perhaps a non-parametric test).

```
. tab case, summ(netdist)
(This means tabulate the summary statistics for netdist by case-control status.)

                 | Summary of netdist
            case |      Mean   Std. Dev.      Freq.
        ---------+----------------------------------
               0 |  5.8931393  2.3231356        750
               1 |  5.2892354  2.3020266        150
        ---------+----------------------------------
           Total |  5.7924886  2.3292667        900
```

```
. ttest netdist, by(case)

(This means carry out a 2-sample t-test of netdist comparing cases and controls.)


Two-sample t test with equal variances          0: Number of obs =      750
                                                 1: Number of obs =      150

-------------------------------------------------------------------------------
Variable |     Mean    Std. Err.      t      P>|t|       [95% Conf. Interval]
---------+---------------------------------------------------------------------
       0 |   5.893139   .0848289   69.4709   0.0000      5.726609     6.05967
       1 |   5.289235   .1879597   28.1403   0.0000      4.917825    5.660646
---------+---------------------------------------------------------------------
    diff |   .6039038   .2074755   2.91072   0.0037       .1967106    1.011097
-------------------------------------------------------------------------------
Degrees of freedom: 898

                   Ho: mean(0) - mean(1) = diff = 0

    Ha: diff < 0              Ha: diff ~= 0              Ha: diff > 0
      t =   2.9107              t =   2.9107               t =   2.9107
    P < t =  0.9982          P > |t| =  0.0037          P > t =  0.0018
```

The lower mean distance from the site in cases (5.28 vs. 5.89) is unlikely to have occurred by chance ($p = 0.0037$, two-sided t-test). (Note the tests labelled Ha: diff<0 and Ha: diff>0 are one-sided tests.) The standard deviations are similar in the two groups. We have not examined whether the distributions approximate the normal, but the quite large numbers in the two groups make it unlikely that the sampling distribution of the difference is not reasonably normal. So we now have a statistical test which suggests that cases probably have a smaller average distance from the site.

Activity 3.4

How could you present these data to give a measure of (relative) risk in relation to the site and a test of association?

Feedback

The ratio of cases to controls (the odds) provides a relative measure of risk. However, it cannot be interpreted in its own right as the selection of cases and controls was of course made on the basis of their disease status (and using a predetermined ratio of five cases to each control). Thus, the odds of being a case has no bearing on the proportion of people with cancer in the local population.

However, we can see whether the ratio of cases to controls *changes* in relation to distance from the site. To do this, you would need to define some bands of distance for which the case/control ratios are computed. But deciding how the bands should be defined should not be done *a posteriori*, as the temptation would be to choose those bands which give the risk estimates closest to what you think they 'should' be. Most investigators take n-tiles (tertiles, quartiles, quintiles) or round-number cut-points that give approximately equal numbers of subjects in each group. In the output shown below we used bands that are a compromise between these objectives, with cut-points at 2, 3, 4.9, 6.3, 7.4, 9.3, 9.2 and 10km from the site:

```
. gen ndgp=recode(netdist,2,3,4.9, 6.3, 7.4, 8.3, 9.2,10)

. sort ndgp

. by ndgp:summ netdist
```

(These commands mean generate a set of groups (bands) of netdist with upper boundaries at 2, 3, 4.9, ...km; then sort by these groups; then summarize the statistics of netdist within these groups.)

```
-> ndgp=        2
Variable |      Obs        Mean   Std. Dev.        Min         Max
---------+--------------------------------------------------------
Variable |      Obs        Mean   Std. Dev.        Min         Max
---------+--------------------------------------------------------
 netdist |       77    2.575624   .2991698    2.014389    2.991351

-> ndgp=      4.9
Variable |      Obs        Mean   Std. Dev.        Min         Max
---------+--------------------------------------------------------
 netdist |      268    4.010186   .5425744    3.012888    4.892958

-> ndgp=      6.3
Variable |      Obs        Mean   Std. Dev.        Min         Max
---------+--------------------------------------------------------
 netdist |      148    5.531248   .3846093    4.903982    6.295494

-> ndgp=      7.4
Variable |      Obs        Mean   Std. Dev.        Min         Max
---------+--------------------------------------------------------
 netdist |      103    6.879091   .3325955    6.307556    7.392496

-> ndgp=      8.3
Variable |      Obs        Mean   Std. Dev.        Min         Max
---------+--------------------------------------------------------
 netdist |       82    7.866042   .2764206    7.424296    8.282077

-> ndgp=      9.2
Variable |      Obs        Mean   Std. Dev.        Min         Max
---------+--------------------------------------------------------
 netdist |      116    8.739056   .2642863    8.300086    9.160111

-> ndgp=       10
Variable |      Obs        Mean   Std. Dev.        Min         Max
---------+--------------------------------------------------------
 netdist |       73    9.571881   .2227238    9.222989    9.998764
```

Then tabulating the odds by distance band we get the following data:

```
. tabodds case ndgp
```

(This means generate a table of the odds of being a case by distance group, ndgp.)

```
table of cases (D), controls (H), and odds (D/H)

    ndgp        _D         _H      _odds     ci_low    ci_high
       2         9         24      0.375      0.174      0.807
       3        15         62      0.242      0.138      0.425
     4.9        55        213      0.258      0.192      0.347
     6.3        22        126      0.175      0.111      0.275
     7.4        16         87      0.184      0.108      0.313
     8.3        10         72      0.139      0.072      0.269
     9.2        15        101      0.149      0.086      0.255
      10         8         65      0.123      0.059      0.257

Chisq test for trend =   8.7993 ( 1 df, p =  0.003 )
```

The column of odds is simply the ratio of cases to controls (_D / _H), and the lower and upper confidence intervals for these ratios are given in the final two columns. For the reasons just described, the odds have no significance in themselves, but the fact that they decline with distance from the site indicates that there is a trend of decreasing risk. The Chi-square test for trend (i.e. of change in odds across distance bands) provides a useful test of association and here indicates that the decline in odds is unlikely to be due to chance (p = 0.003).

Regression analysis

These data could also be analysed using a logistic model, the output of which would show *odds ratios* relative to the baseline group. In this case the baseline group is the innermost distance band. Note that the baseline group, which by definition has an odds ratio of 1, is omitted from the output:

```
xi: logistic case i.ndgp
```

(This means perform a logistic analysis of the variable case (coded case=1 for case, and case=0 for control) in relation to the distance group, ndgp.)

```
i.ndgp                    Indgp_1-8    (Indgp_1 for ndgp==2 omitted)

Logit Estimates                                    Number of obs =       900
                                                   chi2(7)       =     10.34
                                                   Prob > chi2   =    0.1699
Log Likelihood = -400.33348                        Pseudo R2     =    0.0128

---------------------------------------------------------------------------
    case | Odds Ratio   Std. Err.      z     P>|z|      [95% Conf. Interval]
---------+-----------------------------------------------------------------
 Indgp_2 |   .6451613    .3131348   -0.903   0.367      .2491897    1.670346
 Indgp_3 |   .6885759    .2885899   -0.890   0.373      .3028334    1.565669
 Indgp_4 |   .4656085    .2114132   -1.684   0.092      .1912168    1.133746
 Indgp_5 |   .4904215    .2335415   -1.496   0.135      .1928513    1.247144
 Indgp_6 |   .3703704    .1912584   -1.923   0.054      .1346102    1.019048
 Indgp_7 |   .3960396    .1896634   -1.934   0.053      .1549167    1.012463
 Indgp_8 |   .3282051    .1777048   -2.058   0.040      .1135709     .9484701
---------------------------------------------------------------------------
```

Thus, the second column ('odds ratios'), are the ratios of the odds for each band relative to the baseline (innermost) band. For Indgp_2, the second band from the site, the odds ratio of 0.65 suggests that the odds of being a case is around 35% lower than in the innermost band. The final two columns show the corresponding confidence intervals.

A test for trend can also be generated by the logistic model by fitting ndgp as a linear term (i.e. as its (group) value rather than as a set of indicators of individual groups). The odds ratio then indicates the average relative change in risk for each band one moved away from the site:

```
. logistic case ndgp
```

(Carry out a logistic analysis of case in relation to ndgp fitted as a linear term.)

```
Logit Estimates                                Number of obs =     900
                                               chi2(1)       =    8.84
                                               Prob > chi2   = 0.0030
Log Likelihood = -401.08681                    Pseudo R2     = 0.0109

------------------------------------------------------------------------
 case |  Odds Ratio   Std. Err.      z     P>|z|    [95% Conf. Interval]
------+-----------------------------------------------------------------
 ndgp |   .8858941    .0363802    -2.950   0.003     .8173842    .9601461
------------------------------------------------------------------------
```

The result (0.89 (95% CI 0.82, 0.96)) provides evidence of decline in risk of about 11 per cent (i.e. 1–0.89 expressed in percentage terms) per band. Alternatively, one might fit the numerical value netdist, in which case the odds ratio is the relative change in risk for each kilometre increase in distance from the site:

```
. logistic case netdist
```

(Fit logistic model of case using netdist, the numerical value of distance in kilometres from the site.)

```
Logit Estimates                                Number of obs =     900
                                               chi2(1)       =    8.51
                                               Prob > chi2   = 0.0035
Log Likelihood = -401.25125                    Pseudo R2     = 0.0105

-------------------------------------------------------------------------
  case  |  Odds Ratio   Std. Err.      z     P>|z|    [95% Conf. Interval]
--------+----------------------------------------------------------------
netdist |   .8925355    .0351975    -2.883   0.004     .8261483    .9642573
-------------------------------------------------------------------------
```

Rather than quantifying the decrease in risk as one moves *away* from the plant, it might be intuitively clearer to express the change in risk as you get closer to the site. This could be done by reversing the order of bands. To do this we generate a new variable called revndgp:

```
. gen revndgp=10-ndgp
. xi:logistic case i.revndgp
```

(Generate a reverse group variable, revndgp, and then perform a logistic analysis of revndgp.)

```
i.revndgp           Irevnd_1-8   (Irevnd_1 for revndgp==0 omitted)

Logit Estimates                                Number of obs =     900
                                               chi2(7)       =   10.34
                                               Prob > chi2   = 0.1699
Log Likelihood = -400.33348                    Pseudo R2     = 0.0128

--------------------------------------------------------------------------
    case |  Odds Ratio   Std. Err.      z     P>|z|    [95% Conf. Interval]
---------+----------------------------------------------------------------
Irevnd_2 |   1.206683    .5620503    0.403   0.687     .4843057    3.006539
Irevnd_3 |   1.128472    .569038     0.240   0.811     .4200182    3.031892
Irevnd_4 |   1.494253    .6918547    0.867   0.386     .6029883    3.702877
Irevnd_5 |   1.418651    .6244892    0.794   0.427     .5986557    3.361816
Irevnd_6 |   2.098005    .8477109    1.834   0.067     .950328     4.631689
Irevnd_7 |   1.965726    .9286471    1.431   0.153     .7787513    4.96189
Irevnd_8 |   3.046875   1.649713     2.058   0.040     1.05433     8.805072
--------------------------------------------------------------------------
```

```
. logistic case revndgp

Logit Estimates                                    Number of obs =      900
                                                   chi2(1)       =     8.84
                                                   Prob > chi2   =   0.0030
Log Likelihood = -401.08681                        Pseudo R2     =   0.0109

---------------------------------------------------------------------------
    case | Odds Ratio   Std. Err.       z     P>|z|    [95% Conf. Interval]
---------+-----------------------------------------------------------------
  revndgp |  1.128803    .0463556    2.950    0.003     1.041508    1.223415
---------------------------------------------------------------------------
```

And we could also generate a reversed variable of distance in kilometres (which takes the value of 0 at 10 km from the site and 10 on top of the site).

```
. gen revnetdi=10-netdist

. logistic case revnetdi
```

(This means generate a new variable equal to 10 - netdist, which therefore gets bigger with proximity to the site, and perform a logistic analysis of case in relation to this new variable.)

```
Logit Estimates                                    Number of obs =      900
                                                   chi2(1)       =     8.51
                                                   Prob > chi2   =   0.0035
Log Likelihood = -401.25125                        Pseudo R2     =   0.0105

---------------------------------------------------------------------------
    case | Odds Ratio   Std. Err.       z     P>|z|    [95% Conf. Interval]
---------+-----------------------------------------------------------------
 revnetdi |  1.120404    .0441836    2.883    0.004     1.037068    1.210436
---------------------------------------------------------------------------
```

The p-values from this and the previous logistic regressions and the trend tests based on distance bands are all the same (0.003). But whereas the odds ratio for netdist represents the odds ratio associated with moving one kilometre further way from the site, for revnetdi it is its reciprocal – the odds ratio associated with moving one kilometre nearer the site. The last results, which show the change in risk per kilometre of distance, indicate that the risk increases by about 12 per cent for every kilometre moved *closer* to the site. This is therefore evidence that the risk of disease is greater in proximity to the industrial plant.

Activity 3.5

What further analyses would you like to do before concluding that there is evidence that cancer risk rises with proximity to the site?

Feedback

The obvious issue to consider is the possibility of confounding, especially by socio-economic status. Socioeconomically disadvantaged people may well live closer to industrial areas because their limited resources give them fewer choices in deciding where to buy or rent a dwelling. But we know also that poorer people tend to have higher rates than average across a broad range of diseases, including most cancers.

Socioeconomic status would thus be a confounding factor associated both with exposure (proximity to the site) and with the outcome of interest (cancer). You could check this in your data:

```
. by case:tab depriv, summ(netdist)

(This means tabulate the average distance (and standard deviation) from the site for each
of the four deprivation groups.)

-> case=           0
          |       Summary of netdist
   depriv |      Mean   Std. Dev.      Freq.
----------+-----------------------------------
        1 |  6.6002917   1.8707853        199
        2 |  5.9670693   2.3377337        192
        3 |  5.8648305   2.4036124        178
        4 |  5.0650789    2.424567        181
----------+-----------------------------------
    Total |  5.8931393   2.3231356        750

-> case=           1
          |       Summary of netdist
   depriv |      Mean   Std. Dev.      Freq.
----------+-----------------------------------
        1 |  5.7700102   2.2523182         26
        2 |  5.9914974   2.1986472         34
        3 |  5.3099149   2.4331147         46
        4 |  4.4408649   2.0594929         44
----------+-----------------------------------
```

```
. tab case depriv, col

(Cross-tabulate deprivation group with case-control status.)

           | depriv
      case |        1         2         3         4 |    Total
-----------+--------------------------------------------+----------
         0 |      199       192       178       181 |      750
           |    88.44     84.96     79.46     80.44 |    83.33
-----------+--------------------------------------------+----------
         1 |       26        34        46        44 |      150
           |    11.56     15.04     20.54     19.56 |    16.67
-----------+--------------------------------------------+----------
     Total |      225       226       224       225 |      900
           |   100.00    100.00    100.00    100.00 |   100.00
```

Thus, deprivation is associated with the outcome – the percentage of cases varies by deprivation group – and with distance from the site – average distance varies by deprivation group. The conditions for confounding have been met. To examine whether the risk of cancer is independently associated with the site, we need to adjust for socioeconomic status. This can be achieved using (multi-variable) logistic regression analysis:

```
. xi:logistic case revnetdi i.depriv

(Perform logistic analysis of case on revnetdi adjusting for deprivation fitted
as four groups.)

i.depriv            Idepri_1-4   (naturally coded; Idepri_1 omitted)

Logit Estimates                              Number of obs =      900
                                             chi2(4)       =    14.44
                                             Prob > chi2   =   0.0060
Log Likelihood = -398.28744                  Pseudo R2     =   0.0178
---------------------------------------------------------------------------
     case | Odds Ratio   Std. Err.       z    P>|z|     [95% Conf. Interval]
----------+----------------------------------------------------------------
 revnetdi |   1.10063    .0444161    2.376   0.018      1.01693    1.191219
 Idepri_2 |  1.282806    .3607757    0.886   0.376      .7392124    2.22614
 Idepri_3 |  1.835628     .493502    2.259   0.024      1.083782    3.109046
 Idepri_4 |  1.598339    .4415018    1.698   0.090      .9301341    2.746581
---------------------------------------------------------------------------
```

On including (adjusting for) deprivation, the odds ratio per kilometre changes from 1.13 to 1.10. So in fact there has been a little confounding. The p-value was more affected, as was the confidence interval. However, there remains quite strong evidence for an association of risk with distance from site (p = 0.02), after adjusting for deprivation. If this were a formal hypothesis test, it would provide evidence that the plant is associated with a higher risk of cancer.

More sophisticated models might look at different risk functions of distance – for example, an exponential or quadratic decline in risk with distance. But the basic model construction would follow the same principles.

Further comments on cluster investigations

Over these first three chapters we have been interested in the investigation of putative disease clusters. In earlier sections we have alluded to the fact that the interpretation of such studies very much depends on the context. We distinguish between two settings:

1 A cluster in search of a causal hypothesis (the context of cluster reports).
2 A causal hypothesis in search of a cluster ('is there clustering of disease around this source of exposure?' – a hypothesis-testing study).

Setting (1) is more common, but is difficult to evaluate and is controversial. Most statistical methods strictly apply only to Setting (2). Setting (1) is an example of what is often referred to as the 'Texas sharp shooter' phenomenon (Figure 3.3). The Texas sharp shooter first shoots . . . then draws the target where most bullets have hit. The epidemiological analogy is the selection of a cluster from the pool of all potential clusters. When someone notices a cluster, they are effectively drawing the target around cases which are close together in space and time. But it is impossible to judge how many similar targets could be drawn in which there is no such aggregation (or cluster) of cases. The number is perhaps very large, and it is

Figure 3.3 The Texas sharp shooter phenomenon

unsurprising therefore that sometimes 'targets' are observed in which the number of cases is high. The difficulty is that the observer cannot really know or test how unusual their particular observation is because they have no measure of the number of potential targets they are ignoring. The cluster might all be due to natural variation.

Rothman (1990) commented on this in his lecture 'A sobering start to the cluster-busters conference'. Because of the difficulties posed by the Texas sharp shooter phenomenon, he concluded that:

- with very few exceptions, there is little scientific or public health purpose to investigate individual disease clusters at all;
- there is likewise very little reason to study overall patterns of disease clustering in space-time; and
- as a consequence no statistical methodologies are needed to refine our study of disease clusters or clustering in general.

This is a polar view, but it reflects a reality that the large majority of investigations of apparent single-site clusters do not ever identify a cause. Most epidemiologists understand that such investigations are likely to lead nowhere, and thus may not be warranted on scientific grounds. In consequence, there is a good argument that the resources which might be devoted to such investigations would be far better spent in other ways.

An alternative perspective is put forward by Neutra (1990), who argues for a more pragmatic approach. In some cases, cluster investigations may be justified by public concern and/or the nature of the apparent cluster. As described in Chapter 1, protocols and guidelines have been developed by agencies such as the US Communicable Disease Center. These generally propose staged assessments, beginning with the rapid assessment of whether there is a *prima facie* case of an excess, followed by review of the reported cases, and then formal epidemiological study. The investigation can be stopped at any stage if the evidence and/or public health context indicate that there is little merit in proceeding further.

Because p-values are of limited use in such studies, greater weight needs to be given to factors such as the specificity of effect, the plausibility of the exposure-disease link(s), dose-response relationships, and the pattern of findings in relation to the timing and level of exposure. Analyses might be undertaken which eliminate the 'index' cases. But with all *post hoc* studies, the interpretation will often be inconclusive. On the other hand, where a hypothesis is advanced without reference to any local data, it is appropriate to apply tests of statistical inference, and p-values can be interpreted in the usual way.

You will recall that the study you have looked at in the first chapter is loosely based on a real-life example, in which a journalist reported an apparent but non-specific increase in risk of cancer near to a pesticide factory. After detailed investigation of the statistics in the vicinity of the plant, the authors of this cluster investigation concluded that the 'study provides limited and inconsistent evidence for a local-ized excess of cancer in the vicinity of the [plant]. At present further investigation does not seem warranted . . .' (Wilkinson *et al.* 1997). Given the circumstance of the original cancer report, this conclusion is what one might have expected before embarking on the study.

Summary

This chapter introduced the statistical analysis of point (case-control) data relating to local environmental exposure. Graphical plots may indicate whether disease risk is higher close to the source of hazard, but formal statistical analyses are also needed. A trend of risk with distance from the site provides a reasonable global test of association, and can be based on tabulation or logistic regression models. Adjustment for socioeconomic confounding is often important because of the association of social disadvantage both with disease risk and residence close to industrial areas.

An important distinction was drawn between hypothesis-testing studies, which can be interpreted in the normal way, and cluster investigations, which are extremely difficult to interpret because there is no satisfactory method to assess the role of chance (Texas sharp shooter phenomenon). Because of this difficulty, some epidemiologists argue that there is seldom any merit in studying a cluster report as the investigation is unlikely to lead to any useful insights. However, the decision about how far to proceed with a cluster enquiry is likely to be dictated by a range of factors, including public concern. Various guidelines have been proposed.

References

Neutra RR (1990). Counterpoint from a cluster buster. *American Journal of Epidemiology* **132**(1): 1–8.

Rothman KJ (1990). A sobering start for the cluster busters' conference. *American Journal of Epidemiology* **132**(1 Suppl): S6–13.

Wilkinson P, Thakrar B *et al.* (1997). Cancer incidence and mortality around the Pan Britannica Industries pesticide factory, Waltham Abbey. *Occupational and Environmental Medicine* **54**(2): 101–7.

Useful websites

Centers for Disease Control, Atlanta, GA: www.phppo.cdc.gov/cdc

Agency for Toxic Substances and Disease Registry (ATSDR): www.atsdr.cdc.gov

National Institute for Occupational Safety and Health (NIOSH): www.cdc.gov/niosh

US Environmental Protection Agency (EPA): www.epa.gov

SECTION 2

Air pollution

4

Air pollution: time-series studies

Overview

For centuries, people have understood that air pollution harms human health. In the UK, the early part of the twentieth century saw an increase in the burning of coal which led to a dramatic rise in levels of smoke and sulphur dioxide. This rise remained unchecked until the famous 1952 London smog episode which was responsible for a two- to threefold increase in mortality and showed beyond doubt that episodes of high air pollution have a detrimental effect on respiratory and cardiovascular health. Since that time, ambient levels of air pollution have decreased due to the Clean Air Acts of 1956 and 1968 and other factors. In the present day, the main source of urban air pollution is from motor vehicles. However, much of the recent epidemiological evidence points to an adverse pollution effect on health even at modest levels observed in many cities today. Most of these epidemiology studies have used time-series methods of analysis to investigate pollution effects. These studies assess any short-term effects of air pollution on health by estimating associations between day-to-day variations in both air pollution levels and in mortality and morbidity counts. Despite the growing evidence from these kinds of studies, questions remain about the mechanisms involved, the effects of chronic exposure, susceptible populations and strategies of amelioration.

In this and the following chapter you will learn about the most common types and sources of modern-day air pollution, and the main epidemiological designs that are used to assess their effects on health. This chapter concentrates on time-series studies.

Learning objectives

By the end of this chapter you should be able to:

- describe the principal epidemiological approaches used to investigate short-term consequences of air pollution exposure
- explain the basic design features of time-series studies for investigating the health effects of environmental exposures
- describe the strengths and weaknesses of these designs
- explain the concept of mortality displacement

Key terms

Mortality displacement (harvesting) The name given to the bringing forward in time by just a few days or weeks of death or other health event by an environmental exposure.

Particulates Particulate matter, aerosols or fine particles of solid or liquid suspended in their air.

Time-series studies The analysis of variation in events, such as daily or weekly counts of deaths or hospital admissions, in relation to exposures measured at similar temporal resolution.

Types and sources of air pollution

A wide range of pollutants exist, but those of chief concern from a health perspective are:

- particles (such as PM_{10})
- sulphur dioxide (SO_2)
- nitrogen oxides (NO_x) including nitrogen dioxide (NO_2)
- ozone (O_3)
- carbon monoxide (CO)
- volatile organic compounds (VOCs) including benzene
- lead (Pb)

Carbon dioxide is quantitatively the most important gas emitted by fossil-fuel burning. It has no direct effects on health, but it does contribute to global warming. Most attention has focused on particle fractions, especially particles of small diameter that can enter the respiratory tract. Evidence of adverse health effects is strongest for particles with a diameter of less than 10 microns (so-called PM_{10}) and less than 2.5 microns ($PM_{2.5}$). $PM_{2.5}$ are respirable, that is small enough to penetrate deep into the lung. Ultrafine particles have a diameter less than 0.1 microns.

Ozone has been shown to have effects on lung function in some subjects, probably through inflammatory/irritant processes, and carbon monoxide by binding to haemoglobin can reduce the oxygen-carrying capacity of the blood which may be of particular importance for people with severe cardio-respiratory limitation. The health impacts of other pollutants are less clear.

Figure 4.1 shows the principal sources of emissions of the main pollutants in the UK in 2001. Ozone is not emitted directly from man-made sources in any significant quantities, but arises from chemical reactions in the atmosphere caused by sunlight. The relationship between concentrations and emissions is complex and influenced by patterns of dispersion, air chemistry and other factors.

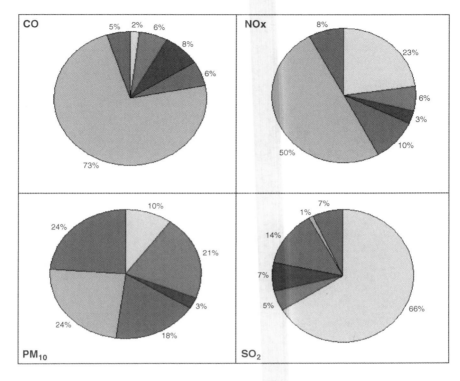

- ☐ Power stations
- ☒ Domestic & commercial
- ■ Refineries, iron & steel
- ☒ Other industry
- ☐ Transport
- ☒ Other (e.g. agriculture, construction)

Figure 4.1 Sources of priority pollutants in the UK in 2001
Source: NETCEN 2005

✎ Activity 4.1

Figure 4.2 shows maps of emissions of SO_2 and NO_x in Britain in 2001. The dark sections show the areas where emission levels are at their highest; in the case of the right-hand map these broadly tend to be around the motorway networks. Using the information in Figure 4.1, which map shows emission levels for SO_2 and which for NO_x? The Air Quality Strategy for England, Scotland, Wales and Northern Ireland identifies the action that needs to be taken at international, national and local level to reduce emissions of air pollution. In particular, it provides a framework which allows relevant parties, such as industry, business and local government to identify the contributions they can make. What kinds of actions that can be made at an individual level would you consider important in helping meet the objectives of the strategy?

Figure 4.2 Emission levels of two pollutants in Great Britain
Source: National Atmospheric Emissions Inventory (www.naei.org.uk/)

Feedback

The map on the right-hand side shows emission levels for NO_x. This is known because the areas of highest concentrations (the dark sections) in this map are occurring mostly in the cities and along the main motorway networks. These are the areas where traffic levels are highest, and as was seen in Figure 4.1, the main source of NO_x in the UK is transport. Several actions can be made at an individual level to reduce emissions. For example, not using cars for short journeys, sharing car journeys with friends and family and having cars serviced regularly.

Types and sources of air pollution

Pollution is also affecting the whole world. The burning of fuel in power stations and oil refineries provides the energy for use in homes and cars. This burning of fuel also pumps out 'greenhouse gases' which cause global warming. In the UK this could mean more floods and storms, hotter summers and wetter winters. Saving energy and resources can help to keep fuel consumption to a minimum.

As has been discussed, the main source of present-day air pollution in the UK is motor vehicles. However, it was a very different picture in the early part of last century when an increase in the burning of coal led to a dramatic rise in levels of smoke and sulphur dioxide. This rise remained unchecked until the famous 1952 London smog episode which was responsible for a two- to threefold increase in

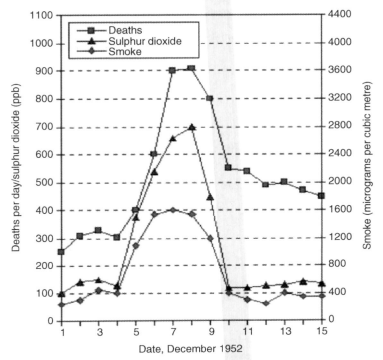

Figure 4.3 Smoke, sulphur dioxide and mortality levels in London during the December 1952 smog episode

Source: Wilkins 1954

mortality and showed beyond doubt that episodes of high air pollution have a detrimental effect on respiratory and cardiovascular health (Ministry of Health 1954). Figure 4.3 shows the peak in mortality during the smog episode coinciding with peaks in smoke and sulphur dioxide levels.

Since that time, pollution produced from the burning of coal has substantially reduced, in part due to the Clean Air Acts passed in 1956 and 1968. Figure 4.4 shows the dramatic reduction in levels in these pollutants over recent decades. Since the beginning of the 1990s, attention has switched to 'newer' pollutants such as PM_{10} and NO_2 due to increases in traffic volume.

The government, the European Community and the World Health Organization set standards and guidelines for levels of air pollution. These are concentrations that are considered to be acceptable in the light of what is known about the effects of each pollutant on health and the environment. A partial summary of current UK objectives is shown in Table 4.1. The table also displays the highest level of each pollutant reached as recorded by the London Bloomsbury monitoring site in 1998. Observed daily PM_{10} levels exceeded the government standard on three occasions during 1998.

Figure 4.5 shows observed daily levels of PM_{10} in central London between 1995 and 2003. Although levels did not breach the government standard of 50 µg/m^3 (broken

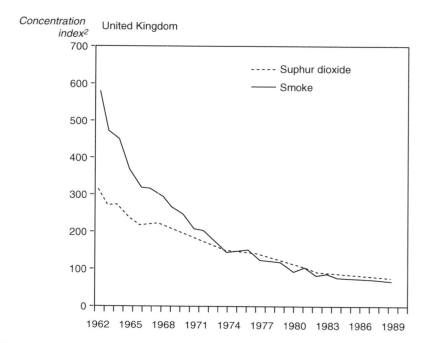

Figure 4.4 Smoke and sulphur dioxide: trends in urban concentrations

Source: Committee on the Medical Effects of Air Pollutants (1995)

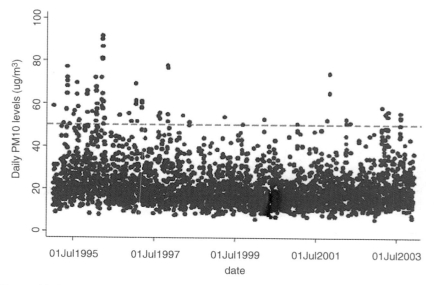

Figure 4.5 Daily levels of PM10 in central London 1995–2003

Table 4.1 Summary of objectives of the UK National Air Quality Strategy and summary statistics of observed levels, Bloomsbury (London), 1998

Pollutant	Standard		Highest level reached in Bloomsbury in 1998
	Objective	Measured as	
Carbon monoxide	10 ppm	Maximum daily running 8-hr mean	2.2 ppm
Nitrogen dioxide	250 ppb	1-hr mean	65 ppb
Ozone	50 ppb	Running 8-hr mean	31 ppb
Particles (PM_{10})	50 µg/m³, not to be exceeded more than 35 times per year	24-hr mean	61 µg/m³
Sulphur dioxide	100 ppb	24-hr mean	36 ppb

ppm = parts per million; ppb = parts per billion; µg/m³ = microgrammes per cubic metre

line) more than 35 times/year in this period, levels do exceed 50 µg/m³ fairly regularly – suggesting that high pollution days do still occur but not very often.

Despite the so-called 'safe' limits, much of the recent experimental and epidemiological evidence points to an adverse pollution effect on health, even at modest levels, observed in many cities today.

Studies of health effects

Over the years, a wide variety of research has been conducted to assess the effects of air pollution on health. The most common designs include:

- laboratory studies (also called chamber studies)
  - humans
  - animal models;
- panel and event studies;
- large population studies
  - time-series
  - geographical comparisons.

Chamber, panel and event studies are designed so that individuals are studied, though they may rely upon aggregate-level exposure information. The two types of population studies – time-series and geographical – are the main epidemiological designs. Geographical studies are covered in the next chapter. Time-series studies are the most common type of study and the main concepts of such designs are discussed below.

Time-series studies

Time-series studies assess the effects of short-term changes in air pollution on health events by estimating associations between day-to-day variations in air pollution on the one hand and mortality or morbidity counts on the other. The data on outcome and exposure (and possibly confounders) for time-series analysis usually comprises daily pollution levels and daily mortality or hospitalization counts for a given area for a number of years. Short-term effects are then estimated using regression analyses of health event count (Y) on pollution level (X), though specific features of the time-series data need to be respected.

These studies are ecological (exposure defined at group level) because the unit of analysis is the area – usually an entire city. However, the temporal nature of time-series studies avoids some of the concerns about confounding in ecological studies. Risk factors that do not change over short durations of time, such as smoking habits, use of gas for cooking, and social class are the same on polluted as unpolluted days. We can say that the design utilizes the population in question as its own control. Similarly, the persons at risk change only slowly over time (births, migration and deaths), and so are not taken into account as 'denominators' in time-series studies. The outcome variable is usually the daily count of the health outcome, not the rate.

Although factors that change little over time do not confound time-series studies, factors that do change in time can do so. For example, if mortality decreases over time, due perhaps to improved diet or reduced deprivation, and air pollution decreases, a spurious 'confounded' association of mortality with air pollution will be found. We are helped here by the focus in time-series studies on acute effects – associations that exist on a short-term basis (that is, over the space of a few days or weeks). We can therefore use statistical methods to 'filter out' long-term trends and fluctuations in mortality, and so exclude confounding by factors operating on such long-term time scales. These long-term fluctuations can be systematic trends over years or seasonal variations repeated over the course of each year (season). Other potential confounding from measurable time-varying factors operating at short time scales, such as temperature, humidity, influenza, day of the week, and public holidays can be controlled by inclusion of appropriate variables in the regression analysis.

Other issues common to time-series studies are:

- *Temporal autocorrelation*. Outcome data on adjacent days may be highly correlated with each other. Special models reduce the tendency to make confidence intervals too narrow, but if autocorrelation remains high on allowing for measured potential confounding variables this indicates potential for residual confounding.
- *Poisson distribution of outcome*. A usual multiple regression model can be used to predict outcome for each day, but because the outcome is a count (or deaths or hospitalizations), the distribution of actual counts around this predicted value is more likely to follow a Poisson than the normal (Gaussian) distribution usually assumed. A special kind of regression called Poisson regression is therefore preferable.
- *Overdispersion*. Counts of health outcome data, though often approximately Poisson distributed, are frequently 'overdispersed', which means that they have more variation than predicted by the Poisson model. This can be allowed for by using a simple modification of the method.
- *Shape of exposure-response function*. This is usually assumed to be linear in the case of air pollution, which allows for easy quantification of effect sizes.
- *Lags*. Pollution on any given day may affect health on that same day, but also the day after, and the day after that, etc. Again an extension of standard methods allow such delayed effects to be investigated.

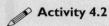 **Activity 4.2**

Read the extract below, which is taken from an air pollution time-series study published by Anderson *et al.* (1996), and then answer the following questions:

1   What are the outcome and explanatory variables of interest?
2   What are the data?
3   What variables were allowed for as potential confounders?
4   Which of the issues common to time-series studies listed as bullet points above are addressed in the extract?

 Air pollution and daily mortality in London: 1987–92

Objective – To investigate whether outdoor air pollution levels in London influence daily mortality. Design – Poisson regression analysis of daily counts of deaths, with adjustment for effects of secular trend, seasonal and other cyclical factors, day of the week, holidays, influenza epidemic, temperature, humidity, and autocorrelation, from April 1987 to March 1992. Pollution variables were particles (black smoke), sulphur dioxide, ozone, and nitrogen dioxide, lagged 0–3 days. Setting – Greater London. Outcome measures – Relative risk of death from all causes (excluding accidents), respiratory disease, and cardiovascular disease. Results – Ozone levels (same day) were associated with a significant increase in all cause, cardiovascular, and respiratory mortality; the effects were greater in the warm seasons (April to September) and were independent of the effects of other pollutants. In the warm season an increase of the eight hour ozone concentration from the 10th to the 90th centile of the seasonal change (7–36 ppb) was associated with an increase of 3.5% (95% confidence interval 1.7 to 5.3), 3.6% (1.04 to 6.1), and 5.4% (0.4 to 10.7) in all cause, cardiovascular, and respiratory mortality respectively. Black smoke concentrations on the previous day were significantly associated with all cause mortality, and this effect was also greater in the warm season and was independent of the effects of other pollutants. For black smoke an increase from the 10th to 90th centile in the warm season (7–19 microg/m3) was associated with an increase of 2.5% (0.9 to 4.1) in all cause mortality. Significant but smaller and less consistent effects were also observed for nitrogen dioxide and sulphur dioxide. Conclusion – Daily variations in air pollution within the range currently occurring in London may have an adverse effect on daily mortality.

 Feedback

1   In this study the particular outcome of interest is mortality – all causes (excluding accidents), respiratory disease and cardiovascular disease. The pollution variables measured were particles (black smoke), sulphur dioxide, ozone and nitrogen dioxide.

2   The data comprise the daily counts of deaths in London over a five-year period. Although not explicitly stated, it can be assumed that the main air pollution exposure variables would also have been recorded for the same time period and at the same daily resolution.

3   The 'design' section specifies that both trend (secular trend) and season (seasonal and other cyclical factors) were adjusted for in the regression model. Other potentially confounding variables included day of the week, holidays, influenza epidemic, temperature

and humidity. Each of these factors were controlled for as they may be related to mortality and also to the exposure variables of interest.

4 Bullet point issues:

- Autocorrelation was controlled for. We are not told the size of residual autocorrelation.
- It is also stated that Poisson regression was used.
- No mention is made in the extract about whether overdispersion was present or allowed for.
- No information is provided on the exposure-response function, although the authors have assumed a linear relationship here since they present their results as an estimated change in mortality for a specified increase in the pollution exposure (in this case from the 10th percentile of the pollution distribution to the 90th percentile).
- The pollution variables were lagged 0–3 days. This means that the effects of each pollutant measure was assessed on mortality on the same day as the day of exposure (lag 0), but also on mortality the day after exposure (lag 1), two days after exposure (lag 2) and three days after (lag 3). This models the effect pollution may have on deaths on the same day as exposure but also if any effects persist up to three days later.

Figure 4.6 shows a time-series of observed and fitted values of mortality (loge) from the above study. Also shown are the residual values obtained from the difference

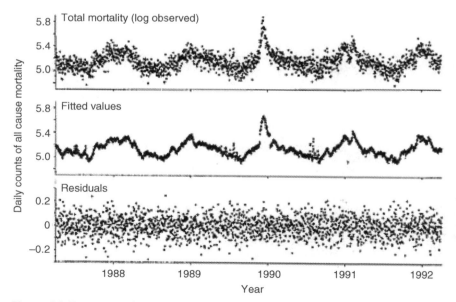

Figure 4.6 Time-series of daily counts of observed and fitted all-cause mortality (loge) in London 1987–92. Residuals are the difference between observed and fitted values

Source: Anderson et al. (1996)

between the observed and fitted values. The fitted values control for potential confounders such as temperature and the strong yearly seasonal pattern observed in the data, but do not adjust for air pollution at this stage. Any associations remaining thereafter between the residuals and the pollutant exposure of interest should, in principle, be free of confounding. (An actual analysis would include the potential confounders and air pollution in the model simultaneously. This approximate procedure is shown to help you understand the logic of the method.)

Activity 4.3

Table 4.2 shows selected results of air pollution effects on mortality from the same study.

1 In this study, which pollutant seems to have the strongest association with mortality, and which cause of deaths is most affected?
2 The estimates presented are displayed as percentage increases, and are derived from the relative risk by subtracting 1 and multiplying by 100. Which of the above results are statistically significant at the 5 per cent level?
3 The estimates presented are for a 10th to 90th centile change in each pollutant. What is the relative risk of all cause mortality associated with a one-unit increase in all-year ozone levels? What can you say about the magnitude of your relative risk?
4 The full table in the paper presents one estimate for each pollutant measure on the single day lag that gave the most significant result. What are the dangers of presenting the results in this selective fashion?

Table 4.2 Percentage increase (95% confidence intervals) in daily all cause, cardiovascular and respiratory mortality associated with increase in pollutant level from 10th to 90th centile. Results are for whole year and for cool and warm seasons separately using the single day lag associated with the largest effect*

Pollutant (10th–90th centile)	All cause	Cardiovascular	Respiratory
Ozone	lag 0	lag 0	lag 0
All year (3–29)	2.43 (1.11 to 3.76)	1.44 (−0.45 to 3.36)	6.03 (2.22 to 9.99)
Cool season (2–22)	0.77 (−0.88 to 2.44)	−1.69 (−3.99 to 0.68)	6.20 (1.67 to 10.94)
Warm season (7–36)	3.48 (1.73 to 5.26)	3.55 (1.04 to 6.13)	5.41 (0.35 to 10.73)
Nitrogen dioxide	lag 1	lag 0	lag 1
All year (24–51)	0.75 (−0.08 to 1.60)	0.62 (−0.58 to 1.84)	−0.92 (−3.22 to 1.33)
Cool season (25–49)	0.46 (−0.44 to 1.36)	−0.11 (−1.38 to 1.17)	−0.25 (−2.54 to 2.10)
Warm season (23–53)	1.45 (−0.25 to 3.17)	2.54 (0.18 to 4.96)	−2.90 (−7.55 to 1.99)
Black smoke	lag 1	lag 1	lag 1
All year (8–23)	1.70 (0.82 to 2.58)	0.58 (−0.68 to 1.85)	0.66 (−1.62 to 2.99)
Cool season (9–26)	1.56 (0.45 to 2.67)	0.13 (−1.46 to 1.74)	0.76 (−2.05 to 3.64)
Warm season (7–19)	2.45 (0.88 to 4.05)	1.87 (−0.34 to 4.13)	0.64 (−3.80 to 5.29)

*Relative risk may be obtained by dividing % increase by 100 and adding one. The natural logarithm of relative risk divided by number of units of air pollution between 10th and 90th centile will result in original regression coefficient from Poisson model.

↻ Feedback

1 Ozone appears to have the strongest association with mortality of those studied. This is demonstrated by the larger relative risks associated with this pollutant than with either NO_2 or black smoke. The specific effects of PM_{10} (which is a subset of black smoke) may have been larger, but were not analysed in this study – daily PM_{10} measures only become routinely available in the UK in the 1990s. Note that as these were from regression coefficients relative risks derived cannot generally be compared across explanatory variables. They can be here, however, because they have all been scaled to represent risk at the 90th relative to the 10th percentile.

2 Since the estimates have been converted from a relative risk into a percentage change, a value of 0 would be expected if there was no effect. Therefore, all estimates where the 95 per cent confidence intervals exclude a value of zero are statistically significant at the 5 per cent level. These are black smoke with all-cause mortality, and ozone with respiratory disease and with all-cause and cardiovascular disease in the warm season and all-year analysis. Note the negative estimates would suggest a reduction in death counts associated with pollution exposure, however all of these negative estimates could have arisen by chance.

3 The paper reports 10th–90th percentile changes in pollution to allow a direct comparison across the pollutants – for example, a one-unit increase in ozone may be very different to a one-unit increase in carbon monoxide levels. In the case of all-year ozone, the percentage change in deaths of 2.43 corresponds to a relative risk of 1.0243 (divide by 100 and add 1). The natural logarithm of this relative risk is 0.024 which is the regression coefficient for a 10th–90th percentile change in pollutant – in this case a range of 26 ppb. So we divide by 26 to obtain the coefficient for a one-unit increase, giving a value of 0.000923. We can then exponentiate this to obtain the relative risk of death associated with a one unit change in ozone. This relative risk is 1.0009 (95 per cent CI 1.0004, 1.0014). It can be seen that the relative risk is very small. In general, short-term effects of air pollution on mortality are small, however the exposure is a ubiquitous one and so population burdens are potentially very large.

4 Selecting and presenting only the most significant results leads to upwardly biased estimates. In an extreme case it could be that pollution had no effect on mortality on all other lag measures tested, however this wouldn't be clear from selective presentation of results in this way. In addition, the large number of outcomes, pollutants and lags being tested means that some results would have been statistically significant purely by chance alone (1 in every 20 if all tests were independent of each other). The results need to be interpreted within the context of this multiple testing.

Similar time-series methods are employed when considering other time-varying factors, such as temperature, as the main exposure of interest (see Chapter 12).

Mortality displacement (harvesting)

One of the difficulties of judging the public health importance of results from time-series studies is the issue known as harvesting. The excess of deaths during pollution episodes may be related to the early deaths of people who already have severe cardio-respiratory disease. In many cases, death may be brought forward only by a day or so; and because the pool of susceptible individuals is thereby depleted, the rise in deaths during the episode may, in theory at least, be followed by a compensating decline in cases. If this short-term acceleration of death accounts for all of the excess deaths during a pollution episode, over the long term, no more deaths would occur than in the absence of air pollution. This is shown schematically in Figure 4.7.

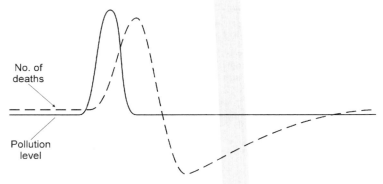

Figure 4.7 Schematic representation of harvesting

Some investigators have sought for but not found the delayed deficit of deaths represented by the dotted line in the figure. Thus, for those studies at least, the deaths associated with recent air pollution do not seem to be displaced by only a few days. However, these methods cannot exclude the possibility that the deaths were displaced rather longer – say a few months.

Time-series studies study acute effects, sometimes called 'triggers', of health events. They suggest that on days of high pollution, deaths, hospital admissions and general practitioner consultations may rise by a few per cent compared with days of low pollution levels. However what is arguably most important, but largely unknown, is the extent to which new disease – more lung cancers, new cases of asthma etc. – is induced by chronic exposure over periods of months and years.

Such chronic effects may arise through a different patho-physiological mechanism from the acute effects, and they cannot be quantified from daily time-series studies which specifically remove long-term trends in disease. Their quantification requires cohort or (less desirable) cross-sectional studies comparing the incidence/prevalence of disease in populations exposed to different annual average levels of pollution. These geographical studies are discussed in the next chapter.

Summary

There are various types of air pollutants, which include the pollutants derived from the burning of fossil fuels by industrial, commercial, domestic and transport-related sources, but also biological materials and dusts. Most of these components have some effects on health, but the epidemiological literature has tended to focus on particle fractions. The health effects of such pollutants may be studied using time-series methods that relate variations in pollution levels (usually measured at daily resolution) to changes in mortality or other health events measured at similar resolution. Such methods have design advantages, but they provide evidence only about acute effects and uncertainties can arise in relation to their public health significance because of the phenomenon of mortality displacement.

References

Anderson HR, de Leon AP *et al.* (1996). Air pollution and daily mortality in London: 1987–92. *BMJ* **312**: 665–9.

Committee on the Medical Effects of Air Pollutants (1995). *Asthma and Outdoor Air Pollution.* London, HMSO.

Ministry of Health (1954). *Mortality and Morbidity During the London Fog of December 1952.* London, HMSO.

National Environmental Technology Centre (NETCEN) (2005). www.swenvo.org.uk/environment/sec4.asp

Wilkins ET (1954). Air pollution aspects of the London fog of December 1952. *Quarterly Journal of the Royal Meteorological Society* **80**: 267–71.

5

Air pollution: geographical studies

Overview

This chapter extends discussion of air pollution epidemiology using extracts from key papers. In the last chapter, you considered time-series studies which provide evidence about the short-term effects of air pollution. We now turn the focus on chronic effects, which require comparisons between populations. In this chapter you will meet the concept of the semi-ecological design, discuss its strengths and weaknesses, and also consider an example of an intervention study.

Learning objectives

By the end of this chapter you should be able to:

- **describe the basic design features of geographical and semi-ecological designs for investigating the long-term health effects of environmental exposures**
- **describe the strengths and weaknesses of such designs**
- **contrast the evidence of geographical and time-series studies, and explain the uncertainties in our knowledge of the health effects of outdoor air pollution**

Key terms

Residual confounding Distortion of the exposure-effect relationship (confounding) that remains after attempted adjustment for the effect of confounding factors.

Semi-ecological design A term often applied to cohort studies of air pollution impacts on health in which exposure is defined at group level (by centrally-located pollution monitor) but data on other risk factors are available at individual level.

Studies of chronic effects of air pollution

In Chapter 4 you were introduced to the principles of time-series studies, the design of which is specifically tailored to assessing acute health effects. They generally entail analysis of the variation in health (mortality, hospital admission, emergency attendance) at daily or weekly resolution, and so focus on exacerbation rather than induction of disease. Their interpretation is also complicated by uncertainty over the degree to which the association between pollution and health outcomes is explained by the harvesting phenomenon.

Studying chronic health effects requires a different design in which pollution exposure and outcome are assessed over the long term. The basic comparison is between populations rather than of the same population over short periods of time.

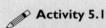

Activity 5.1

Read the extract and study Figure 5.1 relating to the 'six cities' study of Dockery *et al.* (1993).

1 Why do you think its design is sometimes referred to as semi-ecological?
2 What do you think are its particular advantages for air pollution epidemiology?

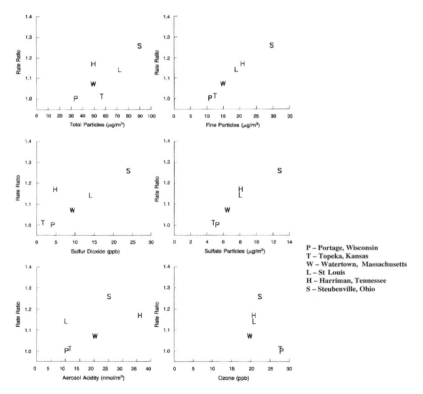

Figure 5.1 Estimated adjusted mortality rate ratios and pollution levels in the six cities. Mean values are shown for air pollution

An association between air pollution and mortality in six US cities

Background Recent studies have reported associations between particulate air pollution and daily mortality rates. Population-based, cross-sectional studies of metropolitan areas in the United States have also found associations between particulate air pollution and annual mortality rates, but these studies have been criticized, in part because they did not directly control for cigarette smoking and other health risks.

Methods In this prospective cohort study, we estimated the effects of air pollution on mortality, while controlling for individual risk factors. Survival analysis, including Cox proportional-hazards regression modeling, was conducted with data from a 14- to 16-year mortality follow-up of 8111 adults in six US cities.

Results Mortality rates were most strongly associated with cigarette smoking. After adjusting for smoking and other risk factors, we observed statistically significant and robust associations between air pollution and mortality. The adjusted mortality-rate ratio for the most polluted of the cities as compared with the least polluted was 1.26 (95 percent confidence interval, 1.08 to 1.47). Air pollution was positively associated with death from lung cancer and cardiopulmonary disease but not with death from other causes considered together. Mortality was most strongly associated with air pollution with fine particulates, including sulfates.

Conclusions Although the effects of other, unmeasured risk factors cannot be excluded with certainty, these results suggest that fine-particulate air pollution, or a more complex pollution mixture associated with fine particulate matter, contributes to excess mortality in certain US cities.

 Feedback

This is often referred to as the 'six cities' study. It has been one of the most influential papers on the health effects of air pollution published in the modern phase of air pollution epidemiological research. It contributed much to the debate about the potential harm of contemporary levels of air pollutants found in cities in North America and other high-income countries.

1 The design is sometimes referred to as semi-ecological because it used air pollution monitoring stations in each city to classify the level of air pollution exposure of all study participants from the same city. Classification of exposure at group level defines the study to be an ecological design.

2 However, unlike most ecological studies, it also gathered individual-level data on non-exposure variables for each of the 8111 participants (a cohort design). This aspect is crucial to the strength of the study, as it allowed more secure comparisons to be made between cities in relation to air pollution effects. With any between-population comparison, the chief concern is whether any observed difference can reliably be attributed to difference in the exposure rather than to some other (confounding) factor(s). That attribution is usually insecure where we are dealing with grouped analysis without data on individual confounding factors. In the six cities study, regression methods (Cox proportional hazards analysis) could be used to compare the mortality experience of the six cities over 14–16-year periods while controlling for principal confounders. The weakness of previously published studies which did not control for individual-level confounders was referred to in the first paragraph of the extract.

Its finding that the adjusted mortality rate for the most polluted city compared with the least was 1.26 (95% CI 1.08, 1.47) provided the first substantive evidence of chronic health effects from ambient pollution levels.

However, it is worth noting that the effective unit of analysis is the city rather than the individual. And having just six cities contributes to uncertainty in interpreting the cause of any differences in health outcome. Nonetheless, the authors point to the specificity of impact on cardio-respiratory outcomes, and from fine particles (rather than other air pollutants, including total particles concentrations). The near perfect straight line of rate ratio vs. fine particle concentration (Figure 5.1) is an illustration of this. This specificity, the individual-level control for principal confounders and the biological plausibility contributed to the strength of evidence for a cause-and-effect association. Its finding of harm from particle pollution was also in keeping with previously published time-series studies, which are methodologically strong.

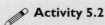 Activity 5.2

Now look at the next extract and Table 5.1, which report a more recent cohort study from the Netherlands (Hoek et al. 2002). What are the similarities and differences from the six cities study?

 Association between mortality and indicators of traffic-related air pollution in the Netherlands: a cohort study

Background Long-term exposure to particulate matter air pollution has been associated with increased cardiopulmonary mortality in the USA. We aimed to assess the relation between traffic-related air pollution and mortality in participants of the Netherlands Cohort study on Diet and Cancer (NLCS), an ongoing study.

Methods We investigated a random sample of 5000 people from the full cohort of the NLCS study (age 55–69 years) from 1986 to 1994. Long-term exposure to traffic-related air pollutants (black smoke and nitrogen dioxide) was estimated for the 1986 home address. Exposure was characterised with the measured regional and urban background concentration and an indicator variable for living near major roads. The association between exposure to air pollution and (cause specific) mortality was assessed with Cox's proportional hazards models, with adjustment for potential confounders.

Findings 489 (11%) of 4492 people with data died during the follow-up period. Cardiopulmonary mortality was associated with living near a major road (relative risk 1.95, 95% CI 1.09–3.52) and, less consistently, with the estimated ambient background concentration (1.34, 0.68–2.64). The relative risk for living near a major road was 1.41 (0.94–2.12) for total deaths. Non-cardiopulmonary, non-lung cancer deaths were unrelated to air pollution (1.03, 0.54–1.96 for living near a major road).

Interpretation Long-term exposure to traffic-related air pollution may shorten life expectancy.

Table 5.1 Risk of cardiopulmonary, non-cardiopulmonary non-lung cancer, and all-cause mortality associated with long-term exposure to traffic related air pollution, NLCS subcohort 1986–94

Model*	Variable	Cardiopulmonary	Non-cardiopulmonary non-lung cancer	All-cause		
				Unadjusted (n = 4466)	Adjusted† (n = 3464)	Adjusted‡ (n = 2788)
1	Black smoke (background)	1.34 (0.68–2.64)	1.15 (0.63–2.10)	1.37 (0.95–1.97)	1.17 (0.76–1.78)	1.04 (0.65–1.64)
	Major road	1.95 (1.09–3.51)	1.03 (0.54–1.96)	1.35 (0.93–1.95)	1.41 (0.94–2.12)	1.53 (1.01–2.33)
2	Black smoke (background and local)	1.71 (1.10–2.67)	1.09 (0.71–1.69)	1.37 (1.06–1.77)	1.32 (0.98–1.78)	1.31 (0.95–1.80)
3	Nitrogen dioxide (background)	1.54 (0.81–2.92)	1.07 (0.61–1.90)	1.37 (0.97–1.94)	1.24 (0.83–1.86)	1.09 (0.70–1.69)
	Major road	1.94 (1.08–3.48)	1.04 (0.54–1.97)	1.34 (0.93–1.95)	1.41 (0.94–2.11)	1.53 (1.01–2.32)
4	Nitrogen dioxide (background and local)	1.81 (0.98–3.34)	1.08 (0.63–1.85)	1.45 (1.05–2.01)	1.36 (0.93–1.98)	1.25 (0.83–1.89)

Values are relative risk (95% CI). Values are calculated for concentration changes from the 5th to the 95th percentile. For black smoke, this was rounded to 10 μg/m³, for NO₂ 30 μg/m³. Adjusted for age, sex, education, Quetelet-index, occupation, active and passive cigarette smoking, and neighbourhood socioeconomic score. *Models 1 and 3 contain the background concentration and an indicator variable for living near a major road. Models 2 and 4 contain an estimate of the home address concentration, by adding to this background concentration a quantitative estimate of living near a major road. Major road is an indicator variable (0/1, 1 indicating living near a major road). †Adjusted for confounders. ‡For individuals living 10 years or longer at their 1986 address, adjusted for above confounders.

ↄ **Feedback**

This European study was broadly similar in its basic design to the US six cities study but with some important differences. Again, it was based on data from a national cohort study in which measurements were made of confounding factors at individual level. All analyses could therefore be adjusted for such factors (e.g. age, sex, education, deprivation index, occupation, active and passive cigarette smoking, and neighbourhood socioeconomic score). Exposure was separately assessed for each member of the cohort using the 1986 residential address based on measured regional and urban background concentrations and a more individualized indicator variable for living near major roads. It was thus based on the basic principle of comparing mortality impacts from long-term (nine-year) exposure to different levels of ambient pollution (black smoke and nitrogen dioxide) while controlling for individual level confounders.

Overall, the background measures of air pollution were not clearly associated with mortality from all causes or from cardiopulmonary or non-cardiopulmonary causes (note that the lower confidence intervals are mostly below 1), though point estimates were all above 1 and substantially higher for cardiopulmonary than for non-cardiopulomonary, non lung-cancer mortality. However, there was evidence of adverse impact on mortality of people who live close to a main road.

Living close to a main road as a determinant raises the obvious question of residual confounding. It would be reasonable to assume that those who live close to a main road are on average more socioeconomically disadvantaged than those who live further away, in which case their poorer mortality experience could be due to residual confounding. However, against this is again the specificity of the increase in risk, which is much greater for cardiopulmonary disease than for non-cardiopulmonary non-lung-cancer mortality.

Geographical vs. time-series studies

The studies described in the last chapter and this emphasize the difference in design and interpretation of time-series and geographical studies (Table 5.2). Their evidence should be taken as being complementary. Time-series studies are methodologically robust as the same population is compared to itself in day-to-day comparisons, so there are no concerns about confounding by individual-level population factors (though confounding could occur from time varying environmental factors such as temperature and influenza). But their evidence relates only

Table 5.2 Comparison of time-series and geographical studies for studying the health effects of outdoor air pollution

Time-Series	Geographical
• Short-term associations	• Long-term exposure effects
• Robust design	• Questions over between-population comparisons
• Repeated evidence of probable causal effects	• Few cohort studies because of time and cost
• Acute effects only	• Provide evidence on disease induction and chronic effects
• Uncertain PH significance	• Clear PH significance

to short-term impacts, relating to exacerbation of disease, and is of uncertain public health significance, especially given the potential for mortality displacement. Geographical studies on the other hand provide evidence which is of clear public health significance and relates to the effects of long-term exposures including disease induction. However, because they rely on comparisons of different populations, their principal weakness is the potential for residual confounding.

An intervention study

The plethora of epidemiological studies about the health effects of outdoor air pollution has provided fairly persuasive evidence. The research focus has therefore started to shift towards mechanisms of action, the activity of particle fractions, issues of vulnerability and intervention studies.

In 1990, the Irish government introduced a ban on the marketing, sale and distribution of bituminous coal within the city of Dublin. A study of this by Clancy *et al.* (2002), examined the change in concentrations of air pollutants and death rates for 72 months before and after the ban, adjusting for weather, season and changes in population structure. It showed that black smoke concentrations were reduced by two-thirds and sulphur dioxide by a third. Death rates were reduced by 287 deaths per year: total non-trauma were reduced by 5.7 per cent, cardiovascular by 10.3 per cent, respiratory by 15.5 per cent, other deaths by 1.7 per cent.

The authors concluded: 'the ban on coal sales within Dublin County Borough led to a substantial decrease in concentration of black smoke particulate air pollution, a reduction of 243 cardiovascular deaths and 116 fewer respiratory deaths per year'. An extract from the paper is reproduced below for you to study. Direct evidence of this kind may help to enhance the case for interventions with policy-makers.

 ### Effect of air-pollution control on death rates in Dublin, Ireland: an intervention study

Background Particulate air pollution episodes have been associated with increased daily death. However, there is little direct evidence that diminished particulate air pollution concentrations would lead to reductions in death rates. We assessed the effect of air pollution controls – ie, the ban on coal sales – on particulate air pollution and death rates in Dublin.

Methods Concentrations of air pollution and directly-standardised non-trauma, respiratory, and cardiovascular death rates were compared for 72 months before and after the ban of coal sales in Dublin. The effect of the ban on age-standardised death rates was estimated with an interrupted time-series analysis, adjusting for weather, respiratory epidemics, and death rates in the rest of Ireland.

Findings Average black smoke concentrations in Dublin declined by 35·6 µg/m^3 (70%) after the ban on coal sales. Adjusted non-trauma death rates decreased by 5·7% (95% CI 4–7, p<0·0001), respiratory deaths by 15·5% (12–19, p<0·0001), and cardiovascular deaths by 10·3% (8–13, p<0·0001). Respiratory and cardiovascular standardised death rates fell coincident with the ban on coal sales. About 116 fewer respiratory deaths and 243 fewer cardiovascular deaths were seen per year in Dublin after the ban.

Interpretation Reductions in respiratory and cardiovascular death rates in Dublin suggest that control of particulate air pollution could substantially diminish daily death. The net benefit of the reduced death rate was greater than predicted from results of previous time-series studies.

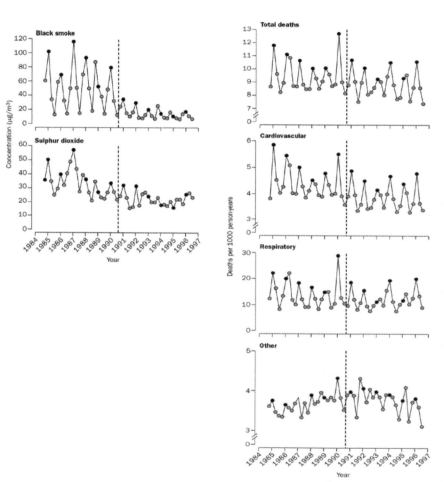

Figure 5.2 Changes in air pollution (black smoke, sulphur dioxide) and in all-cause and cause-specific mortality, Dublin, 1984–97. The ban on coal sales came into effect in 1990

Summary

In this chapter you considered studies of the health effects of long-term exposure to outdoor air pollution. Evidence for such effects comes from comparisons between populations that have been exposed to different pollution levels. The comparisons may be made of populations in different geographical locations or of populations at the same location at different time periods. The main questions of interpretation arise from the need to control for confounding by non-pollution related exposures, such as socioeconomic status, educational level and smoking prevalence. To date, the most robust evidence has come from semi-ecological cohort designs in which such confounding factors have been measured at individual level, even though their costs have limited the number of them. They are important, however, because they provide evidence relating to disease induction (e.g. of lung cancer) and to chronic exposure which time-series studies do not provide. There is also a need for future research to assess the impact of policy interventions aimed at lowering exposures to ambient air pollution.

References

Clancy L, Goodman P *et al.* (2002). Effect of air-pollution control on death rates in Dublin, Ireland: an intervention study. *Lancet* **360**(9341): 1210–14.

Dockery DW, Pope CA *et al.* (1993). An association between air pollution and mortality in six U.S. cities. *New England Journal of Medicine* **329**(24): 1753–9.

Hoek G, Brunekreef B *et al.* (2002). Association between mortality and indicators of traffic-related air pollution in the Netherlands: a cohort study. *Lancet* **360**: 1203–9.

Useful websites

(UK) Committee of the Health Effects of Air Pollutants (COMEAP): www.advisorybodies.doh.gov.uk/comeap/

(US) Environmental Protection Agency: www.epa.gov/

WHO European Centre for Environment and Health, Bonn: www.euro.who.int/eprise/main/who/progs/aiq/home

SECTION 3

Radiation and hazardous waste

6 Ionizing radiation

Overview

Energy may be emitted by certain elements in the form of (invisible) radioactive particles or electromagnetic radiation. Some forms of such radiation can break atomic bonds and is referred to as ionizing radiation. Most ionizing radiation comes from natural sources, though significant doses may be acquired through occupational or medical exposures. The health effects of high doses are well known, and include acute radiation sickness. Evidence regarding low dose exposure is the focus of continuing epidemiological research. Adverse effects may occur through damage to DNA, and include cancer risk, genetic risks and teratogenesis.

In this chapter you will briefly review evidence for health effects of ionizing radiation, and explore the difficulties faced by epidemiological studies. Further information about ionizing radiation and health is available from numerous websites, several of which are listed at the end of this chapter.

Learning objectives

By the end of this chapter you should be able to:

- **describe what is meant by ionizing radiation and how it is measured**
- **describe its sources and the main routes of human exposure**
- **outline the principal categories of potential health effects from exposure to low-dose ionizing radiation**
- **describe the main epidemiological study designs used to investigate adult and childhood cancer in relation to exposure to low-dose ionizing radiation**

Key terms

Becquerel (Bq) The amount of the radioactive material that will have one disintegration in one second.

Electromagnetic spectrum A kind of radiation, including visible light, radio waves, gamma rays and x-rays, in which electric and magnetic fields vary simultaneously.

Genetic effects Effects seen in the offspring of an exposed individual (parent or grandparent) rather than in the individual themselves as a result of damage to genetic material. For an effect to be genetic, exposure must be before conception.

Gray (Gy) The absorbed dose of radiation corresponding to one joule per kilogram.

Ionizing radiation Radiation that is sufficiently energetic to break the bonds that hold molecules together to form ions.

Isotope Each of two or more forms of the same element that contain equal numbers of protons but different numbers of neutrons in their nuclei.

Sievert (Sv) A unit of equivalent dose of radiation which relates the absorbed dose in human tissue to the effective biological damage of the radiation. A milisievert (msv) is one thousandth of a sievert.

Teratogenic effects Abnormalities in the embryo or foetus produced by disturbing maternal homeostasis or by acting directly on the foetus *in utero*.

What is ionizing radiation?

Ionizing radiation is generated by the decay of unstable isotopes of certain elements. It is called 'ionizing' because it is sufficiently energetic to break the bonds that hold molecules together, resulting in charged (ionized) atoms. (This is in contrast to non-ionizing radiation which does not have enough energy to remove electrons from their orbits: see Chapter 7). Some materials are naturally radioactive and some can be made radioactive in a nuclear reactor or particle accelerator. There are two main types of ionizing radiation:

- radioactive particles, including alpha particles (helium nuclei) and beta particles (fast-moving electrons) which ionize matter by direct atomic collisions;
- high-frequency/high-energy electromagnetic radiation such as x-rays and gamma (γ-) rays, which ionize by other types of atomic interaction.

X-rays and γ-rays are part of the electromagnetic spectrum which also includes lower-frequency non-ionizing radiation, such as light, microwaves and radio waves (Figure 6.1).

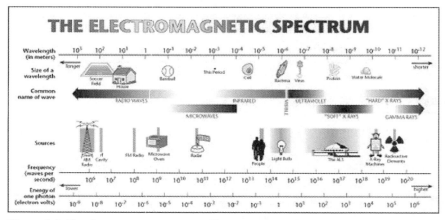

Figure 6.1 The electromagnetic spectrum. The ionizing part of this spectrum includes radiation with wavelengths similar to, or smaller than, the size of a molecule

Source: Microworlds 2005

Alpha particles, which have the largest mass, are readily stopped/absorbed by soft tissue, while beta particles can penetrate soft tissue and paper, but are stopped by sheet aluminium, for example, which mass-less γ-rays are able to penetrate. (Figure 6.2).

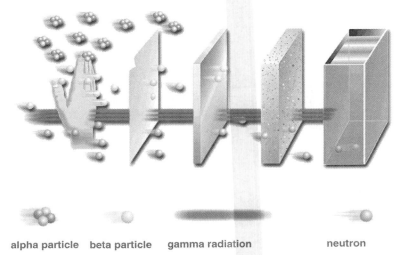

alpha particle   beta particle   gamma radiation   neutron

Figure 6.2 Penetration of the main types of ionizing radiation

Source: Atomic Weapons Establishment 2005

How is it measured?

Ionizing radiation can be measured in terms of the rate of radioactive emissions from a body ('radioactivity') or in terms of the amount of energy absorbed by materials. For health studies, researchers are often concerned with absorption by the human body. The standard international units now used are:

- *Becquerel (Bq).* This is the unit of emission of radioactivity. Specifically, one Bq is the amount of the radioactive material that will have one disintegration in one second. It is named after the French physicist, Antoine-Henri Becquerel (1852–1908), who shared the 1903 Nobel Prize for physics with Marie and Pierre Curie for his discovery of natural radioactivity in uranium salts. The older unit of radioactivity was the Curie (Ci); there are 3.7 million Bq in one Curie.
- *Gray (Gy).* The Gray is a unit of absorbed dose – the amount of energy absorbed by the material (e.g. the human body). Specifically, one Gy is the absorbed dose corresponding to one Joule per kilogram. It is named after Louis H. Gray, an English radiobiologist (1905–65). For those used to using rads (radiation absorbed dose), one Gray is equivalent to 100 rads.
- *Sievert (Sv).* Rolf M. Sievert was a Swedish radiologist (1896–1966). This is a more sophisticated measure of absorbed dose. It is a unit of equivalent dose, which relates the absorbed dose in human tissue to the effective biological damage of the radiation. Not all radiation has the same biological effect, and the absorbed dose is multiplied by a quality factor unique to that type of radiation. Equivalent doses are usually given in milliSieverts (mSv) or microsieverts (μSv). For those

used to using rems, one Sievert is equivalent to 100 rems. Most epidemiological studies (where dose is available) use mSv. A further elaboration of equivalent dose is the effective dose which also takes into account the sensitivity of particular tissues – for example, the lungs, stomach, bone marrow and gonads are more sensitive to ionizing radiation than the bladder, skin or bone cortex. Tissue weighting factors are applied to the equivalent dose to obtain the effective dose, and it is this measure that is used by radiologists to work out appropriate doses in clinical investigations or treatments.

Exposure of human populations

Humans are continuously exposed to radiation from many sources, both natural and artificial. The average yearly dose to the UK population is 2.6 mSv, but there are variations according to place of residence, occupation and medical treatment.

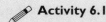 **Activity 6.1**

Think about the possible sources of ionizing radiation. Which do you think are the most important for human exposure? Write down a list and attempt to rank each source and its percentage contribution to overall exposure in the population.

Feedback

Around 85 per cent of exposure to the general population in the UK comes from natural sources, the remaining 15 per cent from man-made sources (Figure 6.3).

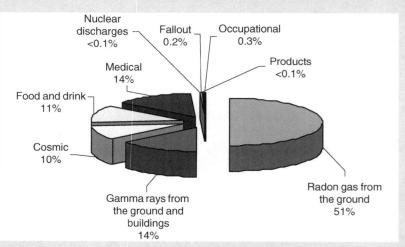

Figure 6.3 Average yearly dose of radiation in the UK

Source: National Radiological Protection Board

Natural sources include γ-radiation emitted by radioactive elements in the earth and building materials. People are irradiated both indoors and out. The average yearly dose from γ-radiation in the ground and buildings is 350 (range 100–1000) μSv; from food it is 300 (range 100–1000) μSv; while the radiation dose from cosmic rays depends on latitude and altitude. The average UK dose from cosmic rays at ground level is 260 (range 200–300) μSv; for passengers on aircraft flying at 10 km altitude, the hourly dose is 5 μSv, and it was around 10 μSv per hour for those who travelled on concord at 15 km altitude. Radon gas comes from uranium that occurs naturally in the ground. Indoors, levels can build up and in some areas (e.g. Cornwall, Devon and Derbyshire) human doses can be very high. The average yearly dose from radon is 1300 μSv, but the range is wide: 300–100,000 μSv.

Man-made sources include:

- *Medical.* X-rays and radioactive materials are used to diagnose disease, and medical radiation is the largest source of man-made exposure to the general public. In the UK the average diagnostic dose is 370 μSv per year. Some patients may get hundreds of times more than this because of cancer treatments.
- *Environmental.* Radioactive materials are discharged to the environment by the nuclear and other industries. The radiation division of the UK Health Protection Agency estimates that the average annual dose to the public is less than 1 μSv, but there is variation by region; some receive up to 200 μSv per year.
- *Radioactive fallout.* The average annual dose to the UK public from weapons testing has declined from 140 μSv in the early 1960s to only 5 in the 1990s. There was a sharp rise in 1986 because of the Chernobyl reactor accident: average annual doses in 1986 reached 20 μSv.
- *Consumer products.* Doses are received from products such as smoke detectors and luminous watches, but the doses are usually small.
- *Occupational exposure (0.3% of exposure in UK).* The most exposed groups are those exposed to natural radiation, especially those exposed to radon in the South West. But about 156,000 people in UK are exposed to radiation in their work. Average annual doses are 1 mSv for the nuclear industry, 0.5 for general radiation workers, and 0.1 mSv for medical radiation workers. Average doses in the nuclear industry in 1991 were half those in 1987.

Health effects of exposure ionizing radiation

The health effects associated with exposure to radiation depend on the degree of exposure. High doses have acute effects. A dose of around 1 Sv is usually sub-lethal, but can give rise to mild radiation sickness, with nausea and vomiting, an early decrease in the number of lymphocytes in the blood and a decrease in all white cells and platelets within two to three weeks. There may be some increase in the long-term risk of leukaemia and solid tumours. A dose of 10 Sv is sufficient to cause severe acute radiation sickness leading to death within 30 days of exposure. At a dose of 100 Sv, coma and death ensues within hours.

Most public health interest concentrates on the effect of low-dose exposures, which are more relevant to occupational or environmental exposures. A low-level dose is defined as being less than 100 mSv. Although this is an arbitrary level, it is a useful number since 100 mSv is (i) the cumulative occupational dose expected to be received by a UK radiation worker after 50 years' work (based on 1988 rates), and (ii) the cumulative dose of the average person in the UK over 45 years from background (natural) radiation. The main health risks that have been studied in relation to low-dose exposure include:

- somatic effects (cancer)
- genetic effects
- teratogenic effects

Evidence on these comes from epidemiological studies of occupational groups (such as nuclear workers), groups of patients who have been exposed during treatment, and survivors of atomic bomb explosions. Much knowledge on the carcinogenic effects of radiation has come from survivors of the A bombs in Hiroshima and Nagasaki, where exposure was acute and generally higher than 100 mSv. Mathematical models generally conclude that the average excess lifetime risk of death from cancer is 0.8 per cent. How far these models can be extrapolated 'downwards' to predict cancer risk following low exposure is still being debated.

Genetic effects

Genetic effects are effects seen in the offspring of an exposed individual rather than in the individual himself or herself. (Effects in the exposed person are termed 'somatic'.) For an effect to be genetic, exposure must be before conception. The range of adverse health outcomes that might be related to pre-conception exposure includes:

- early foetal loss
- childhood cancer
- chromosomal anomalies
- other congenital anomalies
- late foetal loss
- neonatal death
- sex ratio

In a follow-up study by Otake et al. (1990), children born to atomic bomb survivors had increased risk (but not significantly so) of a major anomaly, stillbirth and neonatal death with increasing levels of parental exposure to ionizing radiation.

The two health outcomes that have been investigated most frequently are childhood cancer (diagnosed under 15 years of age) and congenital anomalies (including chromosomal anomalies). Since the mid-1980s there have been numerous investigations of cancer risk in the offspring of persons exposed to low-level ionizing radiation before conception of the child. Particular interest has focused on the possibility that male exposure could result in increased risk of childhood leukaemia in offspring (Gardner et al. 1990). The extract below is taken from a study in the UK (Roman et al. 1999).

Cancer in children of nuclear industry employees: report on children aged under 25 years from nuclear industry family study

Objective: To determine whether children of men and women occupationally exposed to ionising radiation are at increased risk of developing leukaemia or other cancers before their 25th birthday.

Design: Cohort study of children of nuclear industry employees.

Setting: Nuclear establishments operated by the Atomic Energy Authority, Atomic Weapons Establishment, and British Nuclear Fuels.

Subjects: 39 557 children of male employees and 8883 children of female employees.

Main outcome measures: Cancer incidence in offspring reported by parents. Employment and radiation monitoring data (including annual external dose) supplied by the nuclear authorities.

Results: 111 cancers were reported, of which 28 were leukaemia. The estimated standardised incidence ratios for children of male and female employees who were born in 1965 or later were 98 (95% confidence interval 73 to 129) and 96 (50 to 168) for all malignancies and 109 (61 to 180) and 95 (20 to 277) for leukaemia. The leukaemia rate in children whose fathers had accumulated a preconceptual dose of $>/=100$ mSv was 5.8 times that in children conceived before their fathers' employment in the nuclear industry (95% confidence interval 1.3 to 24.8) but this was based on only three exposed cases. Two of these cases were included in the west Cumbrian ('Gardner') case-control study. No significant trends were found between increasing dose and leukaemia.

Conclusions: Cancer in young people is rare, and our results are based on small numbers of events. Overall, the findings suggest that the incidence of cancer and leukaemia among children of nuclear industry employees is similar to that in the general population. The possibility that exposure of fathers to relatively high doses of ionising radiation before their child's conception might be related to an increased risk of leukaemia in their offspring could not be disproved, but this result was based on only three cases, two of which have been previously reported. High conceptual doses are rare, and even if the occupational association were causal, the number of leukaemias involved would be small; in this study of over 46 000 children, fewer than three leukaemias could potentially be attributed to such an exposure.

With regards to anomalies, there have been over 20 surveys into the possible induction of Down's syndrome by low-level irradiation to women. The results are conflicting, but are consistent with the hypothesis that pre-conceptual exposure of women to 20 mGy (for medical reasons) results in a doubling of risk of Down's syndrome. The relationship with other anomalies is not clear. For male pre-conception exposure very few malformation studies have been conducted. Sever *et al.* (1988) examined the risk of congenital anomalies and exposure to radiation before conception among employees in a plutonium and electrical energy plant in USA. No association was found between maternal exposure before conception and the various defects, but for male employees exposure before conception was significantly related to neural tube defects in their offspring. In the UK Nuclear Industry Family Study no association was found between exposures of mothers or fathers before conception to low-dose ionizing radiation at work and risk of chromosomal or non-chromosomal congenital anomalies (Doyle *et al.* 2000).

Teratogenic (*in utero*) effects

Results from the studies of pregnant women exposed during the A bombs of Hiroshima and Nagasaki have shown excess prevalence of microcephaly and mental retardation in their children (Yamazaki and Schull 1990). The gestation at exposure has an important influence on outcome: the period 8–15 weeks gestation has been found to be the most sensitive. Using the same cohort of children, no association was found between dose and the risk of childhood cancer. This is in contrast to previous work by Stewart and Kneale (1970) suggesting an association between maternal exposure to x-rays during pregnancy and risk of childhood cancer death.

Somatic effects: the example of radon

As we saw above, one of the most important sources of exposure to naturally-occurring radiation is from indoor radon. Radon is a naturally occurring, radio-active, noble gas formed as part of the decay chain of uranium–238. It readily diffuses through air and is soluble in water. It is present in small quantities in soil and rock, and can accumulate in enclosed structures, including buildings. The health hazards from radon are well characterized and have been extensively reviewed (BEIR VI 1998).

The hazard derives from the short-lived and chemically reactive isotopes of polonium, lead and bismuth that are its daughter products. When inhaled or formed inside the lungs, these isotopes increase the risk of lung cancer. Epidemiological evidence for this (Boice and Lubin 1997; Brownson and Alavanja 1997) derives mainly from extrapolation from the results of studies of high dose occupational exposures among uranium miners (Hornung 2001), but there have also been case-control (Brownson and Alavanja 1997; Darby *et al.* 1998; Alavanja *et al.* 1999; Field *et al.* 2001; Darby *et al.* 2005) and ecological studies (Stidely and Samet 1993; Darby *et al.* 2001) of residential populations. There is some variation in the risks reported by these studies. The residential studies have not been conclusive in establishing health risks of low-dose radon exposure, but their results are broadly consistent with modelling-based extrapolations from studies of miners (Brownson and Alavanja 1997; Darby *et al.* 1998).

Activity 6.2

Look at the extract below from the 2005 European case-control study of lung cancer risk and indoor radon (Darby *et al.* 2005), and the table of relative risks for different exposure levels.

In some areas of Britain, 30 per cent of homes have radon levels at or above the action level of 200 Bq.m^{-3}. Estimate the proportion of lung cancers that may be attributable to radon in such areas assuming nearly all exposures are in the range 200–400 Bq.m^{-3}. You might find it useful to remind yourself of the formulae for the calculation of population attributable fraction which are summarized in Appendix 3.

Radon in homes and risk of lung cancer: collaborative analysis of individual data from 13 European case-control studies

Objective: To determine the risk of lung cancer associated with exposure at home to the radioactive disintegration products of naturally occurring radon gas.

Design: Collaborative analysis of individual data from 13 case-control studies of residential radon and lung cancer.

Setting: Nine European countries.

Subjects: 7148 cases of lung cancer and 14,208 controls.

Main outcome measures: Relative risks of lung cancer and radon gas concentrations in homes inhabited during the previous 5–34 years measured in becquerels (radon disintegrations per second) per cubic metre (Bq/m3) of household air.

Results: The mean measured radon concentration in homes of people in the control group was 97 Bq/m3, with 11% measuring > 200 and 4% measuring > 400 Bq/m3. For cases of lung cancer the mean concentration was 104 Bq/m3. The risk of lung cancer increased by 8.4% (95% confidence interval 3.0% to 15.8%) per 100 Bq/m3 increase in measured radon (P = 0.0007). This corresponds to an increase of 16% (5% to 31%) per 100 Bq/m3 increase in usual radon – that is, after correction for the dilution caused by random uncertainties in measuring radon concentrations. The dose-response relation seemed to be linear with no threshold and remained significant (P = 0.04) in analyses limited to individuals from homes with measured radon < 200 Bq/m3. The proportionate excess risk did not differ significantly with study, age, sex, or smoking. In the absence of other causes of death, the absolute risks of lung cancer by age 75 years at usual radon concentrations of 0, 100, and 400 Bq/m3 would be about 0.4%, 0.5%, and 0.7%, respectively, for lifelong non-smokers, and about 25 times greater (10%, 12%, and 16%) for cigarette smokers.

Conclusions: Collectively, though not separately, these studies show appreciable hazards from residential radon, particularly for smokers and recent ex-smokers, and indicate that it is responsible for about 2% of all deaths from cancer in Europe.

Table 6.1 Relative risk of lung cancer by radon concentration (Bq/m³) in homes 5–34 years previously

Range of measured values	Mean (Bq/m³)		No of lung cancer cases/controls	Relative risk (95% floated CI)
	Measured values	Estimated usual values		
<25	17	21	566/1474	1.00 (0.87 to 1.15)
25–49	39	42	1999/3905	1.06 (0.98 to 1.15)
50–99	71	69	2618/5033	1.03 (0.96 to 1.10)
100–199	136	119	1296/2247	1.20 (1.08 to 1.32)
200–399	273	236	434/936	1.18 (0.99 to 1.42)
400–799	542	433	169/498	1.43 (1.06 to 1.92)
≥800	1204	678	66/115	2.02 (1.24 to 3.31)
Total	104/97*	90/86*	7148/14 208	–

* Cases/controls. Weighted average for controls, with weights proportional to study specific numbers of cases. Note that as random variation in measured values is approximately logarithmic (so measurement twice as big as usual value is about as likely as measurement half as big as usual value), means of measured values slightly exceed means of estimated usual values.

Source: Darby et al. (2005)

Feedback

To answer this question, you might use one of the standard formulae for population attributable fraction (PAF). For example:

Population Attributable Fraction (PAF) = p(RR − 1)/(p(RR − 1) +1)

where p is the proportion of individuals (homes) exposed, and RR the relative risk. Using the relative risk for the 200–399 Bq.m^{-3} (1.18) and p of 0.30, we obtain:

PAF = 0.3×(1.18 −1) /(0.3(1.18 −1) + 1) = 0.05 or 5%.

In other words, around 5 per cent of lung cancer cases might be attributable to radon in these areas. In fact, because there are higher levels of exposure in some homes, it is reckoned that between 5 and 10 per cent of all lung cancer cases are attributable to radon in Britain as a whole. After smoking, radon and its radioactive progeny are thought to be the most important risk factor for lung cancer in Britain. Other organs may also be targeted by radon through ingestion and skin contact. Malignancies resulting from these exposures may include leukaemia (acute lymphatic leukaemia in children) and skin cancer.

Table 6.2 shows estimates of the absolute risks associated with lifetime radon exposure in the home using two different modelling approaches. The results indicate substantial risks, especially for smokers.

Table 6.2 Estimates of excess annual risks (attributable risk) of lung cancer from lifetime radon exposure in the home

Excess annual risk (cases/100,000) of lung cancer at age 75–79 years from lifetime residential radon exposure by exposure level (Bq m−3), sex and smoking status

| | Exposure-age-concentration model | | | | Exposure-age-duration model | | | |
| | Male | | Female | | Male | | Female | |
Bq m−3	Ever smoked	Never smoked	Ever smoked	Never smoked	Ever smoked	Never smoked	Ever smoked	Never smoked
25	121	29	49	11	81	19	33	8
50	241	58	98	23	162	39	66	15
100	476	116	196	46	320	77	131	30
150	704	173	292	68	476	116	196	46
200	926	231	386	91	628	154	259	61
400	1756	457	750	181	1210	307	509	121
800	3170	898	1417	358	2254	607	978	240

Source: based on BBRM report

Public health action to reduce radon-related cancer has two elements: identifying homes with particularly high indoor levels for mitigation measures and implementation of building regulations to minimize radon accumulation in new homes. The UK government action level for radon (200 Bq.m−3 of air) corresponds to a 3 per cent lifetime risk of lung cancer. Above this level the homeowner is advised to take remedial action.

Summary

In this chapter you learnt about ionizing radiation, which has a number of impacts on health resulting from the damage it can cause to biological molecules (especially genetic material). There are various natural sources of ionizing radiation, the most important of which include radon gas from the ground (problematic in some areas of the UK and other countries), gamma radiation from buildings and the ground, and cosmic rays. There are also several man-made sources, including those relating to medicine, the nuclear industry, radioactive fallout and consumer products. The health effects of high-dose exposure are well documented. Most epidemiological studies now concentrate on the health effects of low-dose exposure that may give rise to cancers, genetic and teratogenic effects.

References

Alavanja MC, Lubin JH et al. (1999). Residential radon exposure and risk of lung cancer in Missouri. American Journal of Public Health 89(7): 1042–8.

BEIR (1998). Biological Effects of Ionizing Radiation (BEIR) VI Report: The Health Effects of Exposure to Indoor Radon. Washington, DC, National Academy Press.

Boice JD and Lubin JH (1997). Occupational and environmental radiation and cancer. Cancer Causes and Control 8(3): 309–22.

Brownson R and Alavanja M (1997). Radon, in Steenland K and Savitz DA (Eds). Topics in Environmental Epidemiology. New York, Oxford University Press.

Darby S, Whitley E et al. (1998). Risk of lung cancer associated with residential radon exposure in south-west England: a case-control study. British Journal of Cancer 78(3): 394–408.

Darby S, Deo H et al. (2001). A parallel analysis of individual and ecological data on residential radon and lung cancer in south-west England. Journal of the Royal Statistical Society Series A 164: 193–203.

Darby S, Hill D et al. (2005). Radon in homes and risk of lung cancer: collaborative analysis of individual data from 13 European case-control studies. BMJ 330(7485): 223.

Doyle P, Maconochie N et al. (2000). Fetal death and congenital malformation in babies born to nuclear industry employees: report from the nuclear industry family study. Lancet 356(9238): 1293–9.

Field RW, Steck DJ et al. (2001). The Iowa radon lung cancer study – phase I: residential radon gas exposure and lung cancer. Sci Total Environ 272(1–3): 67–72.

Gardner MJ, Snee MP et al. (1990). Results of case-control study of leukaemia and lymphoma among young people near Sellafield nuclear plant in West Cumbria. BMJ 300(6722): 423–9.

Hornung RW (2001). Health effects in underground uranium miners. Occupational Medicine 16(2): 331–44.

Otake M, Schull WJ et al. (1990). Congenital malformations, stillbirths, and early mortality among the children of atomic bomb survivors: a reanalysis. Radiation Research 122(1): 1–11.

Roman E, Doyle P et al. (1999). Cancer in children of nuclear industry employees: report on children aged under 25 years from nuclear industry family study. BMJ 318(7196): 1443–50.

Sever LE, Gilbert ES et al. (1988). A case-control study of congenital malformations and occupational exposure to low-level ionizing radiation. American Journal of Epidemiology 127(2): 226–42.

Stewart A and Kneale G (1970). Radiation dose effects in relation to obstetric x-rays and childhood cancers. Lancet 1(7658): 1185–8.

Stidely C and Samet J (1993). A review of ecological studies of lung cancer and indoor radon. *Health Physics* **65**: 234–51.

Yamazaki JN and Schull WJ. (1990). Perinatal loss and neurological abnormalities among children of the atomic bomb. Nagasaki and Hiroshima revisited, 1949 to 1989. *Journal of the American Medical Association* **264**(5): 605–9.

Useful websites

Atomic Weapons Establishment (2005). www.awe.co.uk/main site/scientific and technical/Factsheets/URR/

International Agency for Cancer Research (IARC) – monographs on ionizing radiation: monographs.iarc.fr/htdocs/indexes/vol75index.html

Microworlds (2005). www.lbl.gov/Microworlds/ALSTool/EMSpec/EMSpec2.html

World Health Organization, ionizing radiation: www.who.int/ionizing_radiation/en/

UK Health Protection Agency: www.hpa.org.uk/radiation/

US Environmental Protection Agency (EPA): http://www.epa.gov/radiation/

7 Non-ionizing radiation

Overview

Some radiation (e.g. x-rays) can break molecular bonds and hence produce ions. As we saw in the last chapter, this ionizing radiation has well established health effects, even at low exposures. Other radiation (e.g. from power lines and mobile phones) is non-ionizing, so cannot produce health effects by the same mechanisms. Whether typical exposures to this radiation have any health effects has been controversial. Epidemiological studies have been prominent, but interpretation of them raises particular difficulties.

You will start by reviewing briefly what non-ionizing radiation is, and then briefly what is known about its health effects. Following this, you will explore some of the major issues that have arisen in epidemiological studies through two activities.

Learning objectives

By the end of this chapter you should be able to:

- give a brief overview of the evidence, in particular epidemiological, on the health effects of non-ionizing radiation
- describe the principal difficulties in carrying out and interpreting epidemiological studies of non-ionizing radiation
- describe some issues relating to processes by which epidemiological study results are transmitted to the public
- use a checklist to assist in drawing conclusions on whether a relationship observed in an epidemiological study is causal
- correctly interpret and use the terms job-exposure matrix, cumulative exposure and monotonic relationship

Key terms

Cumulative exposure The total of exposure summed over time, usually the multiplication of the level of exposure (for each job an individual has held) by the duration exposure, summed over all jobs/time periods.

Job-exposure matrix A list of job titles each with an estimated exposure linked to it. Typically, exposure measurements are not made of all workers but rather of a sample of workers which is then applied to other workers with the same job titles.

Non-ionizing radiation Radiation which does not cause the disruption of molecular bonds and hence does not form ions.

What is non-ionizing radiation?

- Non-ionizing radiation is electromagnetic radiation that does not have the energy to break molecular bonds, so cannot produce ions.
- The electromagnetic spectrum is extremely wide. At the high frequency end (ultra-violet, x-ray and gamma ray) radiation is ionizing. At the low frequency end (microwave, radio and power-line frequencies) radiation is non-ionizing.
- Most of the health debate is about radio and power frequencies.
- In radio and higher frequency radiation, the electric and magnetic fields always come together.
- In power (50 or 60 hertz) and lower frequency radiation, electric fields and magnetic fields may each be present without the other, so in effect there are two types of radiation. Nevertheless, in the health literature, these fields are often referred to as extremely low frequency electromagnetic radiation, ELF-EMF, or even just EMF.
- Non-ionizing radiation drops off quickly by distance from the source, typically by an inverse square law. Thus doubling distance from source will often bring exposure down to one quarter the nearer value. For a fuller treatment, see the WHO website noted in 'useful websites' below.

Health effects of non-ionizing radiation

- At low frequencies, external electric and magnetic fields induce circulating currents within the body. In virtually all ordinary environments, the levels of induced currents inside the body are small in relation to natural currents in tissue, so that effects are not strongly plausible.
- The main effect of radio-frequency electromagnetic fields is heating of body tissues. High exposures cause clear health effects through this mechanism. Current exposure standards are designed to avoid heating above negligible levels.
- There is no doubt that short-term exposure to very high levels of electro-magnetic (ELF and radio frequencies) fields can be harmful to health. Current public concern focuses on possible long-term health effects caused by exposure to electromagnetic fields at levels below those required to trigger known biological responses through heating.
- Experiments in which animals have been exposed to quite low-level fields have been inconsistent. Some have found biological effects. Others have not. None have found adverse health effects.
- Some epidemiological studies of ELF-EMF have found some health effects, but results are not consistent. The most consistent evidence is for an elevated rate of leukaemia in children in relation to residential exposure.
- There are very few published epidemiologic studies of radio-frequency fields, although many studies of mobile phone users are in progress. The published studies show little evidence of adverse health effects (excluding road traffic injuries in mobile phone users).
- In summary, despite extensive research, to date there is little evidence to conclude that exposure to low-level electromagnetic fields is harmful to human health. However, the studies cannot exclude small risks, which would be hard to detect, or those due to an aspect of the fields not measured.

- The focus of international research is the investigation of possible links between cancer and electromagnetic fields, at power line (ELF) and radio frequencies.

Role of epidemiology in providing evidence on health effects of non-ionizing radiation

Epidemiology of occupationally and environmentally exposed groups of persons has provided most of what evidence there is on health effects. Carrying out and interpreting these studies, and summarizing the evidence from them, has been difficult and controversial.

You will now explore some of the reasons for the controversy. They are important not just in relation to assessing evidence for health risks of NIR, but also because they illustrate some issues that occur across environmental epidemiology.

Activity 7.1 A (nearly) true story

Your colleague, with whom you share an office (Dr U.B. Careful), has recently completed a study of suicide in relation to EMF. While she is away camping (out of contact), a fax arrives from the journal publishing her article, asking her to check and amend if necessary a press release that they have prepared (see below). You find the fax only an hour before the deadline that they specify for a response, without which you assume they will send out the release as it is anyway.

You have no knowledge of the study, but you do have a copy of the abstract (also below). Your training in public health has given you skills by which, even with this limited information, you may be able to improve on the efforts of the journal's publicity officer. Your boss/conscience, who/which you dare not disobey, tells you to do your best to check if the press release corresponds to the information on the front page, and advise the journal if there seem to be inconsistencies, and (if necessary) of the amendments that would remove them.

- write a sentence summarizing your view of the press release
- if you see the need, propose amendments to the release

The covering letter from the journal press officer

SPECIALIST JOURNALS INC

SJI House

Fax transmission sheet

To: **Dr UB Careful**

From: **B Read, Press releases, Specialist Journals Inc**

Re: **Draft of press release on EMF**

Message
Please check the following press release for accuracy and clarity and to ensure that nothing vital has been omitted or misconstrued. Please amend as you see fit, but I would ask you to bear in mind that we need to run it on the basis of a 'story' and that:

- the readership of these releases is extremely broad, so simple language and brevity are preferred
- the intention is to excite interest, not to summarize the entire paper

As we would like to issue this as soon as possible, I would appreciate you sending amended copy by return. If you have any queries, please do not hesitate to call me on the number listed above.

Thank you for your cooperation.
Yours sincerely
B. Read

Is exposure to electromagnetic fields a killer?

A case cohort study of suicide in relation to exposure to electric electromagnetic fields among electrical utility workers

Cumulative exposure to electromagnetic fields does seem to be associated with a potential for suicide, according to a study published in *Environmental Epidemiology*. Researchers estimated the risk of cumulative and current exposure to 60 Hz electromagnetic fields in 217 electrical workers in Metroland. They had been randomly selected from nearly 22 000 such workers, 49 of whom had committed suicide between 1970 and 1988. Cumulative exposure was graded as high, medium, or low; factors known to be associated with a risk of suicide were also assessed from the company records. These included a history of mental disorder, being unmarried, and alcohol use.

An increased risk of suicide was found among the medium exposure group, having accounted for the other risk factors. There was no evidence to suggest that immediate exposure had any effect on the risk of suicide.

The evidence from previous studies, including several carried out in the UK, has been inconclusive, with some finding a positive association and others refuting any such link. It is thought that electromagnetic fields may affect the function of the nervous system and the production of hormones, in particular melatonin. This hormone regulates sleep patterns and circadian rhythm, disruption to which is strongly linked to depression. The authors of the study are quick to point out, however, that the small sample size and the inability to exclude all other possible suicidal factors weaken the argument for a direct cause and effect.

Abstract
Objectives – This case cohort study examines whether there is an association between exposure to electric and magnetic fields and suicide in a population of 21 744 male electrical utility workers from Metroland.

Methods – 49 deaths from suicide were identified between 1970 and 1988 and a subcohort was selected comprising 1% random sample from this cohort as a basis for risk estimation. Cumulative and current exposures to electric fields, magnetic fields, and pulsed electromagnetic fields (as recorded by the POSITRON meter) were estimated for the subcohort and cases through a job exposure matrix. Two versions of each of these six indices were calculated, one based on the arithmetic mean (AM), and one on the geometric mean (GM) of field strengths.

Results – For cumulative exposure, rate ratios (RR) for all three fields showed mostly small non-significant increases in the medium and high exposure groups. The most increased risk was found in the medium exposure group for the GM of the electric field (RR = 2.76, 95% CI 1.15–6.62). The results did not differ after adjustment for socio-economic state, alcohol use, marital state, and mental disorders. There was a little evidence for an association of risk with exposure immediately before the suicide.

Conclusion – Some evidence for an association between suicide and cumulative exposure to the GM of the electric fields was found. This specific index was not initially identified as the most relevant index, but rather emerged afterwards as showing the most positive association with suicide among the 10 indices studied. Thus the evidence from this study for a casual association between exposure to electric fields and suicide is weak. Small sample size (deaths from suicide) and inability to control for all potential confounding factors were the main limitations of this study.

 Feedback

There is one major issue of concern with the press release, with others less critical. The release places a lot of emphasis on one 'statistically significant' elevated risk – in workers with medium vs. low exposure to electric fields, summarized as the geometric mean. This leads to an interpretation that the study was essentially positive. There are two problems with this:

- As stated in the third-from-last sentence of the abstract, 'This specific index [electric field GM] was not initially identified as the most relevant index, but rather emerged afterwards as showing the most positive association with suicide among the 10 indices studied'. This is a major problem in interpretation, and typical of much EMF epidemiology. There were measurements of three field types – electric, magnetic and pulsed electromagnetic. Two summaries were used for each field type – arithmetic and geometric means (the first emphasizes short peaks, the second more sustained moderate exposures). Finally, researchers considered both lifelong cumulative and recent acute exposure. One might thus calculate that they had 3×2×2 = 12 indices. The abstract mentions ten; the missing two are not explained. Because little is known about the mechanism by which EMF might impact on health, the researchers found no reason to prioritize one index over another. The use of multiple exposure indices (sometimes called 'metrics') without prior specification of which is most likely to be related to disease, has implication for interpretation of significance tests. One significance test out of 20 will produce a 'significant' (p<0.05) association by chance alone. To find one out of ten doing so is not very remarkable. This problem may be referred to as 'multiple testing'. Inappropriately strongly positive interpretation may be referred to as 'data-dredging' or 'post hoc reasoning', and is closely related to the Texas sharp shooter logic discussed in Chapter 3 on clusters.
  - The study subjects were divided into three exposure groups: low, medium and high. The elevated risk was found in the medium exposure group. One can guess that if the

high group had also had a substantially elevated exposure, this would have been mentioned. Most exposures will lead to higher risks with high than medium exposure (a monotonic relationship). Thus this finding, even if just one index of exposure was focused on as most plausibly associated with suicide, provides only modest support for a causal association. Alternatively, we can think of the two relative risks calculated for each index (medium vs. low and high vs. low) as implying $2 \times 10 = 20$ relative risks tested – exactly the number from which you would expect one significant by chance.

Thus the degree of emphasis in the press release on the electric field-GM-cumulative-medium relative risk gives a false impression. The abstract could also be criticised for this, though it is possible to deduce the context from it. Should the researchers have investigated so many associations? Why not focus on just one, and avoid the multiple testing problem? Unfortunately, unless based on clear prior evidence for interest in one exposure measure, this suffers from another limitation. Finding no association with one measure says nothing about whether there could be an association with another.

Less critical concerns and comments include:

- The term 'case cohort' is likely to be unfamiliar to most students, and you might reasonably have been concerned that this was a typographical error for 'case control'. In fact, the case cohort design is a rare but legitimate epidemiologic design that is a slight variant of the nested case control design. Details are beyond the scope of this book.
- The third paragraph comments on increased risk 'found among the medium exposure group', without specifying that this was for cumulative exposure.
- There is no information in the abstract from which to tell whether the text in the last paragraph about possible mechanisms was an accurate reflection of the paper.

Some amendment is required to address the critical concern, and perhaps some of the others. There are reasonable differences in how strong the amendments should be, and the extent to which scientific accuracy should be sacrificed in order to make the release simpler and 'excite interest'. Making no compromises on these matters may be criticized as leaving published research inaccessible to all but specialists.

Although some details have been changed, correspondence essentially similar to that described for this exercise occurred in relation to a paper by Baris *et al.* (1996). A first reaction might be to change 'does seem to be associated' to 'does not seem to be associated' in the first sentence, but this is perhaps too negative. These are the changes eventually agreed on:

Cumulative exposure to electromagnetic fields *may* be associated with a potential for suicide, according to a study published in *Environmental Epidemiology*.

Researchers estimated the risk of cumulative and current exposure to 60 Hz electromagnetic fields in 217 electrical workers in Metroland. They had been randomly selected from nearly 22 000 such workers, 49 of whom had committed suicide between 1970 and 1988. Cumulative exposure was graded as high, medium, or low; factors known to be associated with a risk of suicide were also assessed from the company records. These included a history of mental disorder, being unmarried, and alcohol use.

An increased risk of suicide was found among the medium exposure group *in just one of the five indices studied*, having accounted for the other risk factors. There was no evidence to suggest that immediate exposure had any effect on the risk of suicide.

The evidence from previous studies, including several carried out in the UK, has been inconclusive, with some finding a positive association and others refuting any such link. It is thought that electromagnetic fields may affect the function of the nervous system and the production of hormones, in particular melatonin. This hormone regulates sleep patterns and circadian rhythm, disruption to which is strongly linked to depression. The authors of the study are quick to point out, however, that the small sample size and the inability to exclude all other possible suicidal factors weaken the *strength of evidence that could be provided by the study either for or against* a direct cause and effect.

 Activity 7.2

You are asked to give your advice on how strong the evidence is, from a published abstract (below), for a causal relationship between one type of electromagnetic field and lung cancer (Armstrong *et al.* 1994). You may assume that the abstract does not include simple errors of reporting. We will propose a way of structuring thinking about this, but first read the abstract, and jot down your preliminary conclusion.

 Association between pulsed electromagnetic fields and cancer in electrical utility company workers from Quebec and France

Introduction: We have previously reported on the association of 50–60 Hz electric fields, and 50–60 Hz in a nested case-control study of electrical utility workers in Quebec, Ontario, and France. These studies found some suggestive evidence of an association of exposure to 50–60 Hz magnetic fields and one type of leukaemia, but little evidence for any other association. Here we report the association between pulsed electromagnetic field (PEMF) exposures and cancer in the workers from Quebec and France. (Measures of PEMF exposures were not available from Ontario.) PEMF have been found associated with biological effects in some animal studies. The association of PEMF with cancer has not previously been studied.

Methods: Exposures were assessed through a job-exposures matrix based on about 1000 person-weeks of measurements from exposure meters worn by workers. The PEMF channel of the meter was designed to compile the proportion of the time during which the electric field is greater than 200 V/m in the 5 to 20 MHz frequency band. However, this response was not formally calibrated after the design stage, and the meter was subsequently found to respond to lower magnitude fields in some frequencies, including walky-talky transmissions.

Results: Exposures were considerably higher in Quebec than France. The highest exposed occupations were linemen and splicers.

No association was found between PEMFs and cancers previously suspected of association with magnetic fields (leukaemia, other hematopoetic cancers, brain cancer, or melanoma). However, there was a clear association between cumulative exposure to PEMFs and lung cancer, with odds ratio rising to 3.11 (95% CI 1.60–6.04; $p=0.003$) in the highest exposure group (84 cases). (Table 7.1).

Table 7.1 Odds ratios adjusting for socioeconomic status

Exposure group	No. controls	No. cases	OR*	95% CI
1 <median	229	200	1.00	
2 ≥median-	131	116	1.05	0.74–1.49
3 ≥75th-	85	108	1.84	1.16–2.89
4 ≥90th percentile	63	84	3.11	1.60–6.04

This association was largely confined to Quebec, where there was a monotonic exposure response relationship, with odds ratio of 6.67 (2.68–16.57) in the highest exposed group (32 cases). The association was not explained by smoking or other occupational exposures. However, in a crude SMR analysis, lung cancer mortality of the Quebec workers overall was below that of the general population – a partial contradiction of the above results.

Discussion: The magnitude of the association between PEMF exposure and lung cancer constitutes serious evidence for causality. However, several factors limit the strength of this evidence, notably the lack of precision on what the meters measured, the absence of prior evidence for such an association, and the absence of an overall elevated SMR. Nevertheless, given the public health consequences if the association is causal, testing this hypothesis with other data should be a priority.

 Feedback

Before setting out to assess causality, you should have a reasonable grasp of what was done – the study design. You may have been unsure on the following:

- A nested case-control study is one in which cases and controls are selected from a cohort for which details of case occurrence, time of entering the cohort, and time of leaving it (dying) are known. This largely eliminates selection bias, which can be a problem in population or hospital based case-control studies.
- A 'job-exposure matrix' (JEM) is a list of job titles each with an estimated exposure linked to it. An extract from the JEM for this study is shown in Table 7.2.

Table 7.2 Extract from JEM

Job	Exposure to PEMF*
Generator operator	25
Distribution linemen	
Overground	71
Underground	93
Office worker	0

* Units are complex and not given here.

- The exposure measurements were not made on the cases and controls themselves, but on a sample of workers with the same job titles. In this case 100 workers were sampled and each wore the meter for one week of work. This is not ideal, but a practical necessity, especially if some have died at the time of the study.
- Cumulative exposure is a standard notion in occupational epidemiology, but may be new to you. Each worker's work history (list of job titles, with dates of starting and

stopping) is linked to the JEM, the level of exposure for each job title is multiplied by the duration of time the worker had that job title, and these sub-total summed (see Table 7.3).

• 'Monotonic' is a mathematical term meaning that there is a consistent trend (either upwards or downwards) – each OR in the table is higher than the previous one.

Table 7.3 Cumulative exposure figures

Name	Job title	Start date	End date	Level	Cumulative exposure
Bloggs	Generator operator	1/1/1980	31/12/1985	25	$25 \times 5 = 125$
	Underground lineman	1/1/1986	31/12/1987	93	$93 \times 2 = 186$
	Total				$125 + 186 = 311$

Activity 7.3

Return now to focus on assessing evidence for causality. One standard tool to help in this is Bradford Hill's (1965) checklist: consistency, specificity, dose-response etc. (These are sometimes called 'criteria' for causality, though this is criticized on the grounds that few are requirements, and the importance of each may depend on context.) Additionally, you should consider explanations for any association other than causality – confounding, chance and information bias (exposure or outcome mis-classification). If these are unlikely, a causal association is likely. (Selection bias has been omitted because in a nested case-control study this is a minor issue.)

Checklist of considerations for causality

Strength	?
Consistency	?
Specificity	?
Exposure-response	?
Experiment	?

Alternative explanations

'Chance'	?
Confounders	?
Information bias	?

Against each of the items on the checklist, write, +++, ++, +, /, –, – – or – – –, where:

'+++' evidence on this item strongly supports causality
'/' no evidence on this item
'– – –' evidence on this item weighs in strongly against causality etc.

Feedback

There are no 'right' and 'wrong' answers to this exercise. The allocation of '+++' etc. scores is rather subjective, and we have found experienced epidemiologists differing quite a bit, though the main pattern is quite consistent. What is possible to assess from an abstract alone is limited, but a useful exercise nevertheless. Below are the scores allocated by one experienced epidemiologist, with an explanation of reasoning. Do not worry if they are different from your scores. Pay more attention to the explanations. The value of this exercise is to think systematically about each item.

Checklist of considerations for causality (Bradford Hill selected list)

Strength	++
Consistency	/
Specificity	+
Exposure-response	+++
Experiment	+

Alternative explanations

'Chance'	+
Confounders	++
Information bias	++

Explanations

- *Strength (++)*. The OR of 3.11 would usually be considered a strong association. The OR of 6.67 in Quebec is even stronger – suggesting a possible +++ here. The observation of an SMR below one in all workers downgrades this, but to a limited extent, because: (i) this is the entire workforce, over which the highly exposed comprise just 10 per cent, so the excess in that group may be lost; (ii) the 'external' comparison with the total Quebec population is known to be subject to the 'healthy worker effect', because the general population includes persons too sick to work.
- *Consistency (/)*. There have been no other studies of PEMF and cancer, so consistency of these results with them cannot be assessed. The fact that the association was found in Quebec but not France suggests inconsistency within the study, but this can probably be explained by the lower exposures in France.
- *Specificity (+)*. There are two aspects to specificity. Specificity of outcome (this exposure is not associated with many adverse outcomes) – this seems at least partly supported. In as much as we can tell from the abstract, PEMF was not associated with any other cancer type. Specificity of exposure (this outcome is not association with many adverse exposures) – this seems less relevant here. We cannot tell whether lung cancer was associated with lots of other exposures in these workers.
- *Exposure-response (+++)*. Table 7.3 shows a very strong dose-response relationship. That is, not only does the high (>90th percentile) exposure group show an elevated odds ratio, but the next highest (75th–90th) shows a slightly less high, but also statistically significant association. It is not often that epidemiologic data show such strong evidence of a dose-response relationship.
- *Experiment (+)*. The abstract mentions 'some' animal experiments showing evidence for 'biological effects' (*not* adverse health effects).

Alternative explanations

- *'Chance' (+)*. We see that the 95 per cent CI easily excluding the null value (OR = 1) and p-value is quite low. At face value, this suggests that chance is an unlikely explanation. However, we should be careful to consider whether the Texas sharp shooter phenomenon applies here. The paper only reports results from one exposure index (unlike the ten in Activity 7.1), but mentions that in the same study, 50–60 Hz magnetic and electric fields were also studied and results reported elsewhere were largely negative. More importantly, the study did investigate cancers of all types, and the paper acknowledges that lung cancer was not identified *a priori* as particularly likely to be affected by PEMF. Thus, there is some element of the Texas sharp shooter here. A bit arbitrarily, we have designated what would otherwise be '+++' to '+' because of this. The published paper includes a somewhat more formal assessment.

- *Confounders (++).* The abstract states that 'The association was not explained by smoking or other occupational exposures'. Further, confounding – even by smoking – very rarely explains associations as strong as that seen here. Only an extremely strong risk factor very strongly associated with PEMF exposure could do that.
- *Information bias (++).* This subdivides to outcome misclassification and exposure misclassification. Lung cancer is a serious disease, and not usually mis-diagnosed. Most case ascertainment procedures in developed countries would expect to miss only a few cases. Misclassification is likely to be a very minor problem. Exposure misclassification, however, is likely to have been substantial. Current workers were used to assess the exposure of workers (cases and controls) in the past. However, misclassification is very unlikely to have been 'differential' (more or less likely in cases than controls). Non-differential misclassification leads to bias in associations that is almost always towards there being no associ-ation ('towards the null'). In other words, non-differential misclassification obscures associations, but cannot create spurious ones. There is, however, a special issue with exposure misclassification here, as noted in the second para-graph of the abstract. The above conclusions about misclassification remain, but if the meter in fact was measuring something different from the PEMF for which animal studies reported some biological effects, we should remove the '+' from 'experiment' on the checklist. Usually, we would also have to revisit the 'consistency' score, but we had no evidence for that item anyway.

There is no guidance on how to weight the various items on the checklist in com-ing to an overall summary of evidence for causality. There is even more variation among epidemiologists on this process than on making assessments of each item. Only if all items are '+++' or '++', or if many are '– – –' or '– –', is everyone likely to agree.

For this study, there was variation among the researchers involved. Here the important omission is evidence from which to establish consistency with other studies, or plausibility based on animal studies. Also important is the Texas sharp shooter context, which makes chance a much more plausible explanation than would appear at face value. These two features lead me to consider chance, rather than a causal association, as the most likely explanation. Nevertheless, the strength of the association and in particular of evidence for dose-response is remarkable, and led investigators to conclude that evidence of causality was at least sufficient to warrant further investigation, although this has not, to our knowledge, been undertaken.

Summary

A major difficulty with epidemiology of exposure to non-ionizing radiation (ELF-EMF or radio waves) is lack of clarity on what the potential mechanisms are. If many exposure measures and/or outcomes are investigated and a few associations found, it is hard to know whether they were by chance. If the association between a single exposure measure and a single outcome is investigated and found absent, that provides no evidence on other measures and outcomes. Emphasis on one strong or statistically significant association out of many investigated is a variant of the Texas sharp shooter phenomenon discussed previously, also referred to as multiple testing, data dredging or *post hoc* reasoning. It can be useful to use a checklist of items to consider in assessing causality (e.g. Bradford-Hill's, further discussed in Chapter 14), and of alternative explanations (chance, confounding, information and selection bias).

References

Armstrong B, Thériault G *et al.* (1994). The association between exposure to pulsed electro-magnetic fields and cancer in electrical utility company workers from Québec and France. *American Journal of Epidemiology* **139**: 805–20.

Baris D, Armstrong BG *et al.* (1996). Suicide in electrical utility workers in Quebec: a case cohort study. *Occupational and Environmental Medicine* **53**: 17–24.

Bradford Hill A (1965). The environment and disease: association or causation? *Journal of the Royal Society of Medicine* **58**: 295–300.

IEGMP (2000). *Mobile Phones and Health. Report of an Independent Expert Group on Mobile Phones.* Chilton, IEGMP.

International Commission on non-ionising radiation protection (ICNIRP) Standing Committee on Epidemiology (2001). Review of the epidemiologic literature on EMF and health. *Environmental Health Perspective* **109**(suppl 6): 911–33.

Knave B (1994). Electric and magnetic fields and health outcomes – an overview. *Scandinavian Journal of Work and Environmental Health* **20 spec no.**: 78–89.

Useful websites

WHO EMF project: www.who.ch/emf

National Radiological Protection Board (NRPB): www.nrpb.org/index.htm

London Hazards Centre (Activist): www.lhc.org.uk/members/pubs/factsht/52fact.htm

NIEHS: www.niehs.nih.gov/emfrapid/homr.htm

EPA: www.epa.gov/radiation/

See also IEGMP (2000), International Commission on Non-ionizing Radiation Protection (ICNIRP) Standing Committee on Epidemiology (2001) and Knave (1994).

8 | Hazardous waste and congenital anomalies

Overview

In the past 20 years there has been increasing public concern about environmental pollution and birth defects, with widespread media reporting of putative links with sources of exposure such as waste incinerators, pesticides, air pollution and hormone-disrupting chemicals. Prior to the 1940s it was generally believed that human embryos were protected from the external environment by their extra-embryonic/foetal membranes and the mothers' abdominal and uterine walls. Then in 1941, research on rubella and in 1961 the thalidomide scandal demonstrated that therapeutic drugs could also cause congenital anomalies. In this chapter we will look at a study of environmental exposure to hazardous waste landfill sites and the putative association with congenital anomalies in order to explore the particular features of congenital anomaly epidemiology. You will think about how to assess the significance of this study and critically examine the key methodological issues.

Learning objectives

By the end of the chapter you should be able to:

- **describe the main causes of congenital malformations**
- **describe the main principles of teratogenesis**
- **list known risk factors for congenital malformations in the environment**
- **describe the difficulties of carrying out and interpreting epidemiological studies of congenital malformations**
- **critically assess epidemiological studies of environmental risk factors and congenital malformations**

Key terms

Chromosome Structure(s) found in the nucleus of a cell, made of DNA and proteins, that contains genes. Chromosomes usually come in pairs.

Critical period for congenital anomalies The stage of development of an embryo when it is most susceptible to teratogenic effects. Differs for different organs/organ systems.

Congenital anomaly (malformation) Developmental defects present at birth.

Dose response The magnitude of the effect of a given level of exposure to an agent.

Embryological (developmental) window The time period in which a foetus or embryo is most vulnerable to exposure to a teratogenic agent, after which there is less risk of inducing major congenital anomalies.

Genotype The genetic make-up of an individual which may affect susceptibility to a teratogenic agent.

Mutagen An agent which can cause genetic damage to individual cells.

Phenotype The observable characteristics of the individual.

Teratogen An agent which can induce congenital anomalies in a developing foetus.

Congenital anomalies

Birth defects, congenital malformations and congenital anomalies are all terms used to describe developmental defects present at birth (Moore and Persaud 1993). The Latin *congenitus* means 'born with'. The terms for the science of congenital anomalies ('teratology') and the process of induction of malformation ('teratogenesis') stem from the Geek word *teras*, meaning malformation or monstrosity (Schardein 2000).

The prevalence of clinically significant congenital anomalies at birth is around 2 to 3 per cent (Moore and Persaud 1993). However, the reported prevalence of congenital anomalies may vary considerably depending on factors such as the definition used, the inclusion or exclusion of minor anomalies, the time for which babies are followed up after birth and the completeness of reporting. The incidence of severe multiple anomalies is believed to be higher (10–15 per cent) in early embryos but the majority of these abort spontaneously, often early in the pregnancy. Many may not be detected (the woman may not even know she is pregnant) and so true biological incidence of anomalies is impossible to determine. Hence the rate of anomalies at birth is always referred to as the prevalence rate (never the incidence rate).

Generally the causes of malformations (Figure 8.1) are broadly divided into genetic and environmental (meaning non-genetic), although around 20 per cent of major congenital anomalies are thought to be multifactorial (the result of an interaction between environmental and genetic factors). For up to 60 per cent of congenital anomalies the causes are unknown (Moore and Persaud 1993). There are few well-established specific environmental causes of congenital anomalies (Table 8.1).

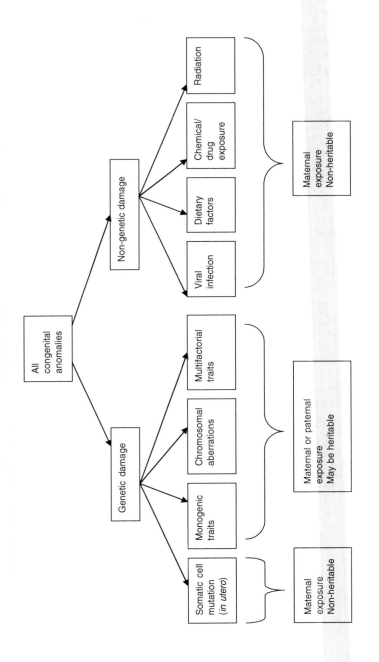

Figure 8.1 Causes of congenital anomalies

Table 8.1 Known environmental causes of congenital anomalies

Medication: Thalidomide, diethylstilbestrol, warfarin, hydantoin, trimethadione, aminopterin methotrexate, streptomycin, tetracycline, valproic acid, isotretinoin, antithyroid drugs, Androgens and high doses of nor-progesterones, Penicillamine, ACE Inhibitors, Carbamazepine, Lithium

Maternal infections: rubella, toxoplasmosis, varicella, venezuelan equine encephalitis, syphilis, cytomegalovirus, herpes simplex virus, human parvovirus B19, Coxsackie A9, Epstein-Barr, Influenza, Fever/hyperthermia

Environmental chemicals: methylmercury, lead, polychlorobiphenyls (PCBs – ingested)

Maternal disorders: insulin-dependent diabetes mellitus, hypo/hyperthyroidism, phenylketonuria, hypertension, autoimmune disorders

Recreational drugs: cocaine, alcohol, LSD, amphetamines, toluene

Key concepts of teratology

- *Critical period* (Figure 8.2): the stage of development of an embryo determines its susceptibility to the teratogen. The most critical period is cell division, cell differentiation and morphogenesis; in later periods minor defects, growth retardation, functional disturbances and physiological defects are more likely. Thus the type and severity of anomalies produced depends on which organs were at their most susceptible at the time of exposure.
- *Genotype of embryo*: there may be differences in the response to a teratogen as a result of the genetic make-up of the individual exposed, for example in babies exposed to phenytoin (a drug used in the treatment of epilepsy and known to be teratogenic). Between 5 and 10 per cent of exposed foetuses develop a phenytoin syndrome with a recognizable spectrum of malformations, 33 per cent only have some of the congenital anomalies and over half are completely unaffected.
- *Dose response and timing*: the response to a teratogen differs as a result of the level of dose of the teratogen and the timing of exposure. Results from a rodent experiment (Figure 8.3) illustrate this. A low dose in the critical period will give rise to a greater degree of insult than a higher dose later in gestation.

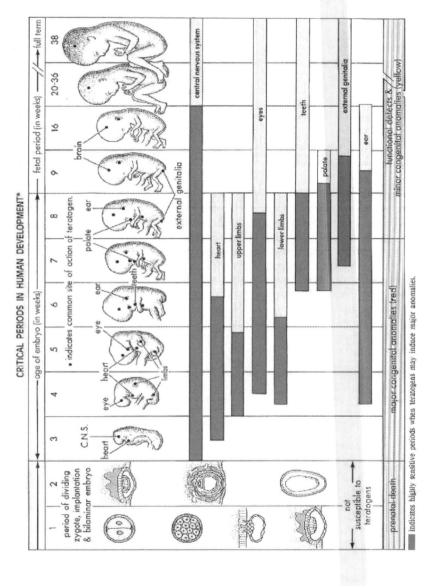

Figure 8.2 Critical periods in human development (darker bar is when major congenital anomalies may be formed, lighter bar is when teratogenic insult may give rise to minor anomalies or functional deficits)

Source: Moore and Persaud (1993)

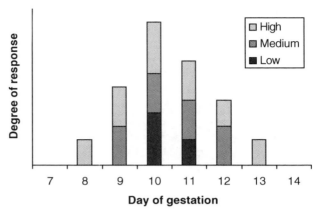

Figure 8.3 Dose response and timing
Source: adapted from Wilson (1973)

Case study: Eurohazcon

Waste disposal by landfill is a cause for environmental concern. Communities close to waste-disposal sites are concerned about the potential health risk of the sites, and may link local 'clusters' of adverse health outcomes to exposure to chemicals from nearby sites. However, even with a random spatial pattern of adverse health outcomes, localized clusters will occur, and distinction of these random clusters from those in which there is a common underlying local cause is difficult. It is desirable to move beyond *post hoc* study of clusters, to study of waste-disposal sites specified *a priori*. The Eurohazcon study (extract below) is a collaborative European study of the risk of congenital anomaly among people living near hazardous waste landfill sites.

  Activity 8.1

Read the following abstract and answer the following questions.

1 What issues need to be considered in deciding how likely it is that hazardous waste might cause human birth defects?
2 Couldn't this excess of congenital anomalies be caused by something else in the local environment?

  Risk of congenital anomalies near hazardous waste landfill sites in Europe: The Eurohazcon study

Background The potential hazards to health of waste disposal sites are a subject of public health concern. The EUROHAZCON study is a multicentric case-control study of the risk of congenital anomaly associated with residence near hazardous waste land-fill sites in Europe. We report here the results concerning non-chromosomal congenital anomalies.

Methods We used data from seven regional registries of congenital anomaly in five countries. We studied 1089 live births, stillbirths, and terminations of pregnancy with non-chromosomal anomalies and 2366 control births without malformation, whose mothers resided within 7 km of a landfill; 21 hazardous waste landfill sites were included. Distance of residence from the nearest waste site was measured, and residence within 3 km designated the 'proximate zone' of most likely exposure to teratogens.

Results Residence within 3 km of a site was associated with a significantly raised risk of congenital anomaly: pooled odds ratio 1.33 (95%CI 1.11–1.59), adjusted for maternal age and socio-economic status. There was a fairly consistent decrease in risk with distance away from the sites. A statistically significant raised odds ratio for residence within 3 km was found for neural tube defects (OR=1.86, 95%CI 1.24–2.79), malformations of the cardiac septa (OR=1.49, 95%CI 1.09–2.04), and anomalies of great arteries and veins (OR=1.81, 95% CI 1.02–3.20). Odds ratios of borderline statistical significance were found for tracheo-oesophageal anomalies (OR=2.25, 95%CI 0.96–5.26), hypospadias (OR=1.96, 95% CI 0.98–3.92), and gastroschisis (OR=3.19, 95% CI 0.95–10.77). There was little evidence of differences between landfill sites in the level of associated risk but power to detect such differences was low.

Interpretation This study shows a raised risk of congenital anomaly in babies whose mothers live close to landfill sites that handle hazardous chemical wastes, although there is a need for further investigation of whether the association of raised risk of congenital anomaly and residence near landfill sites is a causal one. Apparent differences between malformation subgroups should be interpreted cautiously.

 Feedback

1 The likelihood of such chemicals being teratogenic in humans depends on:

- *Dose level*: environmental exposures are usually low (except in unusual circumstances such as industrial accidents) compared to the high levels seen in toxicology. For residents around hazardous waste sites, dose level is very hard to estimate – we will consider this further below. Further, there is some evidence for a threshold of effect in teratogenicity – demonstrated in Figure 8.4.
- *Route of exposure*: different routes of exposure (e.g. ingestion versus subcutaneous injection) have been shown to result in different responses even within the same animal species.
- *Interspecies extrapolation*: response may vary between species and even within different strains of the same animal – for example, the characteristic limb defects that were seen in humans after thalidomide exposure were not seen in any animal species tested.

2 Landfill sites are one possible source of exposure to teratogenic chemicals in the environment. The study gives no information about other possible sources, such as industries which may be causing pollution, however there is little evidence linking other environmental exposures with birth defects. Even if the effect found was due to general pollution or another specific source of pollution, rather than the waste sites, this would still be a cause for concern. Another possibility is that the local population may work in local industries and therefore be occupationally exposed to chemicals at far higher levels than environmental exposures from the waste site. The fact that many sites were

studied, rather than just one, makes the 'other environmental hazard' hypothesis less likely. One landfill site may be close to another environmental hazard, but it seems unlikely that such a pattern would be repeated across the sites of this study.

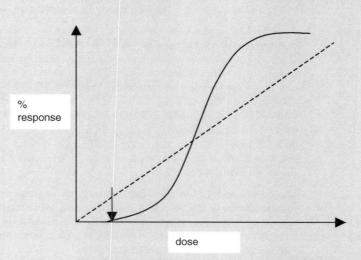

Figure 8.4 Dose response curve for a teratogen (curved line) compared with a carcinogen (dashed line). Arrow shows the threshold dose below which no teratogenic effect is seen

Accurate case definition and completeness of case ascertainment are particularly important for environmental studies of congenital anomalies as variability in case finding or diagnosis may give rise to spurious geographical patterns. Multiple sources of information are generally necessary to ascertain the full range of congenital anomalies, and to overcome inevitable gaps in notification or records.

Terminations have recently been widely used as a result of pre-natal diagnosis of a defect. Thus many pregnancies that would previously have resulted in the birth of a baby with a defect now result in a termination. If such terminations vary geographically they could result in spurious patterns of anomaly prevalence. Terminations due to pre-natal diagnosis are therefore included to avoid such bias. This largely but not entirely resolves the problem: the inclusion of terminations may also result in some babies being counted in the prevalence who might not otherwise have survived to 20 weeks, and if they had been spontaneously aborted would not have been included in prevalence figures.

Inclusion of defects found in terminations not as a result of pre-natal diagnosis is more complex. The detection and reporting of early foetal deaths may vary, as has been mentioned, and the accuracy with which a diagnosis can be made is lower (early foetal deaths are much less likely to be examined by a pathologist, for example). Hence it is usual to include only terminations after a certain gestational age (20 weeks is commonly used) when it is unlikely that a spontaneous abortion will go undetected, and when diagnosis is more likely to be accurate. The study of pre-conceptional mutagenic effects and post-conceptional teratogenic effects requires different study design and interpretation. Thus it is common to exclude chromosomal and other known familial (genetic) conditions from a study of post-conceptional environmental exposures. Other known environmental conditions such as maternal alcoholism or maternal infections (which would be unlikely to be related to chemical exposure) were also excluded.

It is possible that specific exposures may result in very specific anomalies. However, evidence is too limited (because few studies have been carried out and/or because results have been inconsistent) to formulate specific hypotheses about which chemicals cause which birth defects. There is much discussion about how to divide congenital anomalies into meaningful groups (so-called lumping versus splitting). On the one hand, dividing into smaller groups reduces statistical power (thus causing the variability in statistical significance seen in the different anomaly sub-groups in Table 8.2). However, on the other hand, it is also possible that the anomalies in the broad groups are aetiologically heterogeneous, and that the chemical exposure may affect one of the anomalies in a group but not others, so that overall the group does not show an excess. A further argument for dividing into groups is that as there are no known human teratogens that have been found to affect only one organ system, it may be the pattern of the different anomalies which provides more insight into aetiology than increases in the rate of single anomalies or anomaly groups.

Table 8.2 Odds ratios for living within 3 km of a hazardous waste landfill site – selected sub-groups of malformations

Malformation group	N	OR	95% CI
Neural tube defects	130	1.86	1.24–2.79
Hydrocephaly	32	1.06	0.44–2.59
Other central nervous system defects	23	1.03	0.36–2.94
Malformations of cardiac chambers and connections	45	0.91	0.42–1.97
Malformations of cardiac septa	248	1.49	1.09–2.04
Malformations of valves and other heart malformations	109	1.17	0.73–1.88
Anomalies of great arteries and veins	63	1.81	1.02–3.20
Cleft palate	38	1.63	0.77–3.41
Cleft lip with or without cleft palate	72	1.18	0.66–2.12
Tracheo-oesophageal fistula, oesophageal atresia and stenosis	25	2.25	0.96–5.26
Digestive system and upper alimentary tract	59	0.98	0.49–1.93
Atresia and stenosis of rectum and anal canal	20	1.02	0.33–3.15
Hypospadias	45	1.96	0.98–3.92
External genitalia (female + indeterminate)	10	0.89	0.18–4.53
Renal anomalies	75	1.30	0.73–2.31
Urinary tract anomalies	69	1.14	0.62–2.11
Limb reduction defects	41	1.27	0.61–2.62
Exomphalos	12	0.26	0.03–2.19
Gastroschisis	13	3.19	0.95–10.77
Skin and other integument anomalies	30	1.92	0.78–4.73
Syndromes, presumed de-novo mutations	29	1.48	0.63–3.49
Multiply malformed cases	84	1.21	0.71–2.06

Source: Dolk et al. (1998)

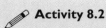 **Activity 8.2**

In this study, exposure was determined by distance of residence from the site. Only landfill sites which had been in operation for at least five years were included. Figure 8.5 shows the odds ratio for congenital anomalies with increasing distance from the site in six distance bands.

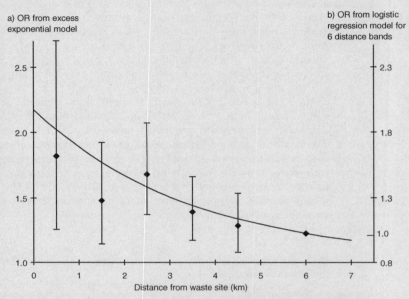

Figure 8.5 Odds ratio (OR) for congenital anomalies with distance from landfill sites

Source: Dolk *et al.* (1998)

Different scales are necessary for the logistic regression and exponential excess model because of the different baseline of the two models. Thus the diamond at 6 km represents the 5–7 km baseline (OR=1, right-hand scale) for the logistic regression of 6 bands, but the solid curve at this point represents estimated risk at 6 km from the site relative to risk infinitely far from the site (OR=1.22, left-hand scale).

1 Do you think distance from the site is an appropriate measure? What alternative measures could be used?
2 Why did the investigators exclude sites which had not been in operation for at least five years?
3 How would you interpret Figure 8.5?

⟲ Feedback

1 Distance from the site is a surrogate measure of exposure. The use of surrogate exposure measurements can lead to misclassification of exposure, usually random, which can in turn lead to an underestimation of the real relative risk. The true pattern of exposure may not depend just on distance from the site. Here knowledge about the main route of exposure is crucial – for example, if the main exposure is via landfill gases

vented to the air then wind speed and direction would be important in determining the true exposure zone, whereas if the main route of exposure was via contaminated groundwater then the position, direction of flow and use of the groundwater sources would be important. Alternative ways to measure exposure such as biological sampling are problematic in this study as the chemical composition of any substances leaking from the landfill is not known and hence it is difficult to decide which chemicals should be sampled.

2 Sites which had been in operation for less than five years were excluded as discussion with the environmental experts suggested that at least five years was required for substantial leakage from a landfill site to occur. The timing of exposure is particularly important for teratogenic exposures since each developmental event occurs in only hours or days, and it is rare for major anomalies to be caused by exposure after the first trimester (as discussed earlier). It should be noted however that for preconceptional mutagenic effects or for chemicals which may have a long half life in the body (e.g. persistent organic pollutants such as PCBs or dioxin) exposure may have occurred well before the study pregnancy.

3 Figure 8.5 shows a decreasing risk with increased distance from the site. The presence of a dose-response relationship strengthens the evidence for causality in any epidemiological study; however, confidence intervals were wide. Although not shown on the figure, in fact the trend of decreasing risk with distance from site was statistically significant (p<0.05).

As with all epidemiological studies it is important to eliminate risk factors which may vary in the same way as the exposure under study and hence act as confounders. Maternal age is an important potential confounder as it is well known that the risk of chromosomal anomalies increases with increasing maternal age. Other maternal factors include parity (the number of pregnancies prior to the index case), social class, lifestyle factors (smoking, alcohol, drug use), maternal disease status (diabetes, epilepsy, fever – related not only to the disease itself but also to the drugs taken to treat the disease) and ethnicity (related to consanguinity, social class, genetic factors etc.). The potential for bias in the selection of cases and controls must also be considered.

Activity 8.3

Is it possible that local concern and knowledge about the waste sites could have affected the study results?

Feedback

Although there is no reason for you to know this, there is little evidence that socio-economic status is associated with congenital anomalies, with the exception of some specific defects (e.g. neural tube defects). However, risk factors (smoking, alcohol) which may be associated with congenital anomalies may also be more prevalent in deprived populations. This study adjusted for socioeconomic status which resulted in very little change in the odds ratio estimates, therefore socioeconomic status does not appear to explain the excess risk found. Residual socioeconomic confounding is possible since it is difficult to measure SES accurately, specially in a study that uses data from different countries.

Increased local awareness might result in more accurate reporting of anomalies in the areas and hospitals close to the sites compared with more distant areas, thus leading to bias. In this study this is unlikely because the data was from routine sources which collect data in a standardized way throughout the whole region. It is possible that anxiety about exposure might cause migration of women planning pregnancy, or women who are pregnant, away from the site – this type of migration is most likely to be random (i.e. the same for case and control mothers – 'non-differential') and would therefore lead to an underestimation of the true relative risk.

 Activity 8.4

As an environmental epidemiologist you may be called on to present your view of a study in as clear terms as possible. Answer the following question which was frequently asked of the investigators by journalists when the results were first published:

based on the Eurohazcon study, if you were planning a pregnancy and lived near a landfill site, would you move?

 Feedback

Arguably, all policy decisions, personal or otherwise, in the face of uncertainty, require value as well as scientific judgements. How much value do you put on the inconvenience of moving? How much value on a defective relative to a non-defective birth? We cannot answer these questions as experts, but we can clarify the scientific input to them.

The study provides some evidence for a risk, but a long way short of proof of one. One epidemiological study is never enough to establish a causal relationship. Other studies have been few. Some have also suggested increases in risk of congenital anomalies near waste sites, but there is no clear pattern from these studies. In order to draw firmer conclusions about causality it would be necessary to do more research, especially to establish exposure pathways and dose.

Further, the results from this study relate to a set of 21 hazardous waste sites. We do not know whether the results can be generalized to all landfill sites. The hazard posed by sites is likely to vary enormously between sites, depending on their age, size, waste types, management and engineering practices, geology etc.

Summary

In this chapter you were introduced to the epidemiology of congenital anomalies and considered the interpretation of a study which found an elevated risk of congenital anomalies around hazardous waste landfill sites. You have considered questions of interpretation similar to those frequently asked by reporters writing news articles. In particular the final question is one which can rarely be avoided – the general public and the media usually want to know what the bottom line is for a study of this kind and although one can endlessly debate the potential sources of error in any epidemiological study, ultimately we must be able to give a final assessment in which such sources of error are weighed against the overall quality of the research and the context of previous research findings.

References

Dolk H, Vrijheid M *et al.* (1998). Risk of congenital anomalies near hazardous-waste landfill sites in Europe: the EUROHAZCON study. *Lancet* **352**: 423–7.

Gregg NM (1941). Congenital cataracts following German measles in the mother. *Trans Ophthalmol soc Aust* **3**: 35–46.

Moore KL and Persaud TVN (1993). *The Developing Human: Clinically Oriented Embryology*. Philadelphia, W.B. Sanders.

Schardein JL (2000). *Chemically Induced Birth Defects*. New York, Dekker.

Wilson JG (1973). *Environment and Birth Defects*. New York, Academic Press.

SECTION 4

Water and health

9 Water and health: a world water crisis?

Overview

This and the following chapter consider the relationship between water and health, beginning with the water cycle and issues relating to the availability and extraction of water for human and agricultural use. You will consider the health implications relating to the water shortage, which arise from the impact of industrialization, population growth, climate change and the global scale of the health burdens arising from inadequate access to clean water and sanitation.

Learning objectives

By the end of this chapter you should be able to:

- **describe the current global 'water crisis' and its origins**
- **identify the potential consequences for health of a future water shortage**
- **describe the global health burdens associated with inadequate access to clean water and sanitation**

Key terms

Disability adjusted life year (DALY) A measure of health based not only on the length of a person's life but also their level of ability (or disability).

Water scarcity Not enough water to supply all users' needs.

Water security A situation of reliable and secure access to water over time. It does not equate to constant quantity of supply as much as predictability, which enables measures to be taken in times of scarcity to avoid stress.

Water shortage The situation where levels of available water do not meet defined minimum requirements.

Water stress The symptomatic consequence of scarcity which may manifest itself as decline in service levels, crop failures, food insecurity etc. This term is analogous to the common use of the term 'drought'.

Sources of fresh water

Although 70 per cent of the earth's surface is covered by water, only 3 per cent is freshwater and therefore fit for human consumption. The bulk of freshwater is locked away in the ice caps of Antarctica and Greenland and in deep underground aquifers, and therefore not readily accessible for human populations. The main sources of freshwater for human use are rivers, lakes, soil moisture and relatively shallow groundwater basins (UNDP 2002). An appreciation of the hydrological cycle (Figure 9.1) is critical to understanding how these various water sources are replenished. This replenishment is largely dependent on evaporation from the world's oceans.

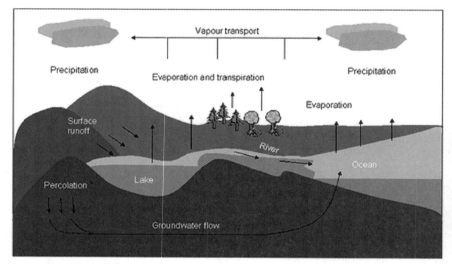

Figure 9.1 The hydrological cycle

Activity 9.1

Consider the sources of freshwater and its relationship with health. List as many activities as you can that are dependent on water in the community/country in which you live. For each one, note how these activities might affect health, either positively or negatively.

Feedback

Water is essential for many aspects of human health. Without a regular supply of freshwater, of sufficient quantity and quality, human life would not last more than a few days. Many of our basic needs (e.g. drinking and food production) are met by water resources, and we use water for a wide range of activities. You may have listed some of the following as important uses of water in your own daily life, and the life of your community/country:

- drinking
- cooking
- personal hygiene
- irrigation (e.g. growing crops)
- industrial and manufacturing processes
- generation of electricity
- leisure (e.g. swimming, sailing)

While this list is not exhaustive, it does cover some of the main uses for water in most communities. If your list contained different items, then you should consider how these might affect (either positively or negatively) human health.

Water is important not just for basic needs, but is also used for producing many of the goods that we use in our daily lives (e.g. electricity, food processing, clothing, cars). There is no doubt that water brings many benefits to human health, and general well-being. However, it is important to remember that water also has the potential to adversely affect health. For example, groundwater used for drinking may be 'naturally' contaminated, as is the case in Bangladesh, where more than 35 million people are exposed to high levels of arsenic (UNESCO-WWAP 2003). Irrigated crop production has improved the nutritional status of many communities, especially in sub-Saharan Africa and Asia. However, these large areas of relatively stagnant water provide ideal breeding habitats for mosquitoes (vectors for diseases such as malaria and yellow fever), and snail intermediate hosts responsible for the transmission of schistosomiasis.

Extremes of precipitation (i.e. above- or below-average rainfall) can also lead to adverse health effects. For instance, too much rainfall can result in flooding and landslides, which in turn can lead to deaths, injuries and other adverse health outcomes. Equally, floods can be beneficial by washing away mosquito habitats. On the other hand, too little rainfall (e.g. drought) can lead to failure of the food harvest, and lack of water for basic sanitation and personal hygiene can result in infections such as trachoma, scabies and intestinal helminth infections.

Water crisis?

Although water is a widespread and renewable resource, increasing demand for it and the pollution of water courses contribute to what is sometimes described as a 'global water crisis'. The 2003 *World Water Development Report* identified the problem as a 'crisis of water governance' (UNESCO-WWAP 2003).

Demand for water is driven by several factors, including population growth and improvement in living standards, with consequent increase in *per capita* consumption. On a global basis we currently abstract for human use around 8 per cent of the total annual renewable freshwater, 26 per cent of annual evapo-transpiration and 54 per cent of accessible run-off (UNESCO/WWAP 2003). Within these global figures there are important regional variations, and many areas of water shortage and scarcity. It is estimated that approximately 40 per cent of the world's population face some level of water shortage. Underground water reserves in many countries are being used faster than they are replenished.

Under the north China plain, which produces nearly 40 per cent of the Chinese grain harvest, the fall in the water table has averaged 1.5 metres a year. Regions under the most water pressure include China's Yellow River basin, the Middle East and the Aral Sea region of Central Asia. Most of the water from these sources is used for irrigation and industry rather than household use. In this context, the inefficiency in water use is a key factor, especially in the area of food production. Approximately 70 per cent of the world's exploited freshwater resource base is diverted to agriculture, and yet only 30 per cent of this diverted water is used by crops and plants.

The availability of freshwater relative to water withdrawals can be an important constraint on development. Unless changes are made, it is estimated that, within the next two decades, the use of water by humans will increase by about 40 per cent and demand will outstrip available supplies. The proportion of the world's population that will be subject to water stress (i.e. where withdrawals of water are more than 20 per cent of the renewable supply) is projected to increase to two-thirds by 2025. Climate change contributes to this increase, as rainfall within many tropical and subtropical regions is expected to decline and to be more erratic; climate change may account for 20 per cent of the increase in water scarcity.

These concerns are reflected in the terms 'water stress' and 'water scarcity', which have become part of the discourse on water resources, and are formal indicators used to describe the pressures placed on water resources at a regional and national level.

This crisis narrative is reflected in much of the literature on water and health, and the following quotations (all three are cited in Mehta 2000) give a flavour of this narrative:

> Many of the wars of this century were about oil, but wars of the next century will be about water.
>
> (Serageldin in Cooper 1995)

> The next war in the Middle East will be over water, not politics.
>
> (Boutros Boutris-Ghali in Butts 1997)

> Not all water-resources disputes will lead to violent conflict [but] in certain regions of the world, such as the Middle East and southern Asia, water is an increasingly scarce resource . . . In these regions, the probability of violence, due to at least in part to water disputes, is increasing.
>
> (Gleick 1993)

This crisis narrative has been continued in the *WWDR*, which states: 'We are in the midst of a water crisis that has many faces . . . The water crisis that exists is set to worsen despite continuing debate over the very existence of such a crisis' (UNESCO-WWAP 2003).

Indicators of water availability

Indicators are important sources of information for academics, policy-makers and numerous other individuals and institutions. In general, indicators are used as a tool to assess how much progress (or lack of progress) is being made in reaching a certain goal or target. They may also be used as a basis to make projections about the future.

It is unlikely that any indicator will give a precise measure – as is indicative from the name, they can only provide an indication of what progress is being made. One issue debated in the literature is the role of broad regional and national indicators of water stress/scarcity, an example of which is the Falkenmark Indicator (UN-HABITAT 2003) – a measure of renewable water resources *per capita* per year. When the indicator is applied, a value of less than 1700 cubic metres *per capita* per year is taken to indicate water stress, and a value of less than 1000 cubic metres *per capita* is taken to indicate severe water stress (or water scarcity). From a supply-demand balancing perspective, this indicator implicitly assumes that demand is directly proportional to population.

In reviewing the Falkenmark indicator a recent UN-HABITAT report suggested that it can 'help draw attention to certain types of water issues, provide heuristic tools through which these issues can be better understood, and create a useful frame-work within which to situate more detailed understandings of specific problems in particular places' (UN-HABITAT 2003). However, the same report advised that it is also important to recognize that the indicator is used to indicate water stress/scarcity primarily at a regional and national scale, and that this can 'create mis-understandings about the actual nature of water issues, and support misguided actions'.

The Falkenmark indicator contains a number of inadequacies, which should be borne in mind when applying the indicator at a national scale. For example, if the indicator suggests water stress at the national scale, this is likely to hide intra-national variations. Also, there are often seasonal variations in the amount of water available, and as the indicator relies on averages this can lead to misinterpretation about stress/scarcity status. There are further questions concerning accessibility and economic capacity. For instance, two countries (one rich, the other poor) may be water stressed according to the Falkenmark indicator, but the poor country, with relatively inaccessible resources, is likely to find it much more difficult than a rich country to improve access to water resources.

Water and waterborne disease

The growing problem of water scarcity adds to the continuing blight of inadequate access to clean water for much of the world's population. There are estimated to be 1.1 billion people who do not have access to an improved water supply and 2.4 billion who do not have adequate sanitation. A lack of access to clean water remains one of the greatest threats to global health, and arguably the largest of the environmental risks to health. A ministerial declaration at the Second World Water Forum in The Hague in 2000 set out a number of challenges as the basis of future action on water (UNESCO-WWAP 2003):

- to reduce by 2015 by one half the proportion of people without access to hygienic sanitation facilities;
- to reduce by 2015 by one half the proportion of people without sustainable access to adequate quantities of affordable and safe water;
- to provide water, sanitation and hygiene for all by 2015.

There are intricate linkages between water and health. In terms of cause-effect relationships there are a number of transmission routes through which water adversely affects human health, and it is important that you are aware of these.

Table 9.1 summarizes a classification system for water- and excreta-related infections. Infection through waterborne transmission of a pathogenic agent a pathogen occurs when humans drink water infected with such as *Vibrio cholerae*. It is important to note that 'all water-borne diseases can also be transmitted by any route which permits faecal material to pass into the mouth (a "faecal-oral" route). Thus, cholera may be spread by various faecal-oral routes, for instance via contaminated food' (Cairncross and Feachem 1993).

Water-washed transmission is associated with conditions of poor domestic and personal hygiene, and there are three types of water-washed disease. The first includes infections of the intestinal tract, such as cholera and dysentery; the second, infections of the eyes or skin (e.g. scabies); and the third, infections carried by lice (e.g. louse-borne typhus). Only the first of these three types of water-washed disease can be transmitted by the faecal-oral route, and thus appears in Category 1 in Table 9.1. The second and third types are considered as being strictly water-washed, and thus appear in Category 2.

Infection by helminths (parasitic worms) can be either water-based (Category 3) or soil-based (Category 4). Water-based infection (e.g. schistosomiasis) results from infection by helminths, which rely on an aquatic intermediate host to complete their life cycle. Soil-transmitted helminths 'are not immediately infective, but first require a period of development in favourable conditions, usually in moist soil' (Cairncross and Feachem 1993). Both Categories 3 and 4 are associated with conditions of poor sanitation and hygiene, as infective eggs are passed in either human urine or faeces.

Water-related insect vectors are the fifth category in Table 9.1. Mosquitoes are the primary disease vector associated with this category, and their primary habitat is in, or close to, water bodies. These bodies of water can vary from flowing rivers and streams to lakes and other relatively stagnant pools of water. The primary vector-borne diseases of concern here are dengue, malaria and several of the arboviruses (e.g. St Louis encephalitis).

In their annual *World Health Report* (WHR), the World Health Organization (WHO) provide detailed information on the global burden of disease (GBD). Table 9.2, which is adapted from the *WHR* 2004, summarizes what data are available for the various water-related health outcomes included in the GBD assessment. The water-related health outcomes listed are responsible for 6 per cent of global deaths and 9.2 per cent of global disability-adjusted life years (DALYs). If drowning is excluded, the figures are 5.6 per cent and 8.6 per cent respectively.

The burden of water-associated ill health is particularly acute in low-income economies, where many communities do not have adequate access to basic

Table 9.1 Environmental classification system for water- and excreta-related infections

Category	Infection (pathogenic agent)	
1 Faecal-oral (waterborne or water-washed)	*Diarrhoeas and dysentries*	*Enteric fevers*
	Amoebic dysentery (P)	Typhoid (B)
	Balantidiasis (P)	Paratyphoid (B)
	Campylobacter enteritis (B)	Poliomyelitis (V)
	Cholera (B)	Hepatitis A (V)
	Cryptosporidiosis (P)	
	E. coli diarrhoea (B)	
	Giardiasis (P)	
	Rotavirus diarrhoea (V)	
	Salmonellosis (B)	
	Shigellosis (bacillary dysentery) (B)	
	Yersiniosis (B)	
2 Water-washed:		
(a) skin and eye infections	Infectious skin diseases (M)	
(b) other	Infectious eye diseases (M)	
	Louse-borne typhus (R)	
	Louse-borne relapsing fever (S)	
3 Water-based:		
(a) penetrating skin	Schistosomiasis (H)	
(b) ingested	Guinea worm (H)	
	Clonorchiasis (H)	
	Diphyllobothriasis (H)	
	Fasciolopsiasis (H)	
	Paragonimiasis (H)	
	Others (H)	
4 Soil-transmitted helminths	Ascariasis (roundworm) (H)	
	Trichuriasis (whipworm) (H)	
	Hookworm (H)	
	Strongyloidasis (H)	
5 Water-related insect vector:		
(a) biting near water	Sleeping sickness (P)	
(b) breeding in water	Filariasis (H)	
	Malaria (P)	
	River blindness (H)	
	Mosquito-borne viruses	
	Yellow fever (V)	
	Dengue (V)	
	Others (V)	

Type of pathogenic agent: B = bacterium; H = helminth; P = protozoan; M = miscellaneous; R = rickettsia; S = spirochaete; V = virus.

Source: adapted from Cairncross and Feachem (1993)

infrastructure such as water and sanitation facilities. Prüss *et al.* (2002) estimated that on a global scale the disease burden from water, sanitation and hygiene was equivalent to 4 per cent of all deaths, and to 5.7 per cent of the total disease burden in DALYs. Although this estimate accounted for diarrhoeal diseases, schistosomiasis, trachoma, ascariasis, trichuriasis and hookworm disease, a number of

Table 9.2 Some water-associated diseases by cause and sex: estimates for 2002

	Deaths (in thousands)						Burden of disease (in thousands)					
	Both sexes		Males		Females		Both sexes		Males		Females	
	Number	%	Number	%	Number	%	Number	%	Number	%	Number	%
Total burden of disease (000s of DALYs)							1,490,126	100	772,912	100	717,213	100
Total deaths (000s)	57,029	100	29,891	100	27,138	100						
All infectious and parasitic disease	10,904	19.1	5,795	19.4	5,109	18.8	350,333	23.5	179,307	23.2	171,025	23.8
Diarrhoeal diseases	1,798	3.2	939	3.1	859	3.2	61,966	4.2	32,353	4.2	29,614	4.1
Malaria	1,272	2.2	607	2.0	665	2.5	46,486	3.1	22,243	2.9	24,242	3.4
Trypanosomiasis	48	0.1	31	0.1	17	0.1	1,525	0.1	966	0.1	559	0.1
Schistosomiasis	15	0.0	10	0.0	5	0.0	1,702	0.1	1,020	0.1	681	0.1
Lymphatic filariasis	0	0.0	0	0.0	0	0.0	5,777	0.4	4,413	0.6	1,364	0.2
Onchocerciasis	0	0.0	0	0.0	0	0.0	484	0.0	280	0.0	204	0.0
Dengue	19	0.0	8	0.0	10	0.0	616	0.0	279	0.0	337	0.0
Japanese encephalitis	14	0.0	7	0.0	7	0.0	709	0.0	338	0.0	371	0.1
Trachoma	0	0.0	0	0.0	0	0.0	2,329	0.2	597	0.1	1,732	0.2
Intestinal nematode infections	12	0.0	6	0.0	6	0.0	2,951	0.2	1,490	0.2	1,461	0.2
Ascariasis	3	0.0	1	0.0	2	0.0	1,817	0.1	910	0.1	907	0.1
Trichuriasis	3	0.0	2	0.0	1	0.0	1,006	0.1	519	0.1	488	0.1
Hookworm infection	3	0.0	2	0.0	1	0.0	59	0.0	31	0.0	27	0.0
Drowning	382	0.7	262	0.9	120	0.4	10,840	0.7	7,458	1.0	3,382	0.5

Source: adapted from WHO (2004)

other water- and sanitation-related diseases were not evaluated, and the results are therefore likely to represent conservative estimates of the scale of the burden of disease resulting from water, sanitation and hygiene.

Interventions to reduce diarrhoeal illness in less developed countries

In industrialized countries diarrhoeal disease is generally not considered to be a major public health priority. Most people make a full recovery after taking some medication and a few days of rest. However, the situation in lower-income countries is very different. Many households do not have access to an adequate and safe supply of water, and also lack basic sanitation and hygiene facilities. These conditions often lead to increased transmission of infectious diseases (see Table 9.1), and result in major adverse effects on human health. Every year diarrhoeal disease causes in excess of 1.7 million deaths (Table 9.2), and most of these occur among children less than 5 years old in low-income countries. Many of these deaths are easily preventable through provision of a regular and adequate water supply, provision of sanitation facilities and adoption of good personal hygiene practices. Numerous studies have attempted to quantify the relative contribution that each of these interventions (i.e. hygiene promotion, sanitation and water supply) can make to reduce the burden of diarrhoeal illness, and have provided mixed results. Fewtrell *et al.* (2005) conducted a meta-analysis of these various interventions (Figure 9.2) and an extract below, along with a figure, shows the results.

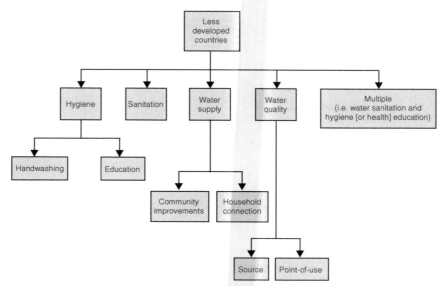

Figure 9.2 Classification of water, sanitation and hygiene interventions
Source: Fewtrell *et al.* (2005)

Water, sanitation, and hygiene interventions to reduce diarrhoea in less developed countries: a systematic review and meta-analysis

Many studies have reported the results of interventions to reduce illness through improvements in drinking water, sanitation facilities, and hygiene practices in less developed countries. There has, however, been no formal systematic review and meta-analysis comparing the evidence of the relative effectiveness of these interventions. We developed a comprehensive search strategy designed to identify all peer-reviewed articles, in any language, that presented water, sanitation, or hygiene interventions. We examined only those articles with specific measurement of diarrhoea morbidity as a health outcome in non-outbreak conditions. We screened the titles and, where necessary, the abstracts of 2120 publications. 46 studies were judged to contain relevant evidence and were reviewed in detail. Data were extracted from these studies and pooled by meta-analysis to provide summary estimates of the effectiveness of each type of intervention. All of the interventions studied were found to reduce significantly the risks of diarrhoeal illness. Most of the interventions had a similar degree of impact on diarrhoeal illness, with the relative risk estimates from the overall meta-analyses ranging between 0·63 and 0·75. The results generally agree with those from previous reviews, but water quality interventions (point-of-use water treatment) were found to be more effective than previously thought, and multiple interventions (consisting of combined water, sanitation, and hygiene measures) were not more effective than interventions with a single focus. There is some evidence of publication bias in the findings from the hygiene and water treatment interventions.

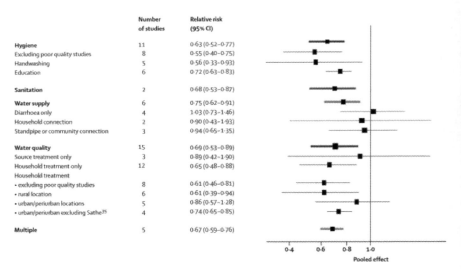

Figure 9.3 Summary of meta-analysis

2 What do these figures tell us about the effectiveness of different forms of intervention?

3 What are the implications of these results for public health policy in relation to the prevention of diarrhoea in less developed countries?

 Feedback

1 The relative risk of 0.63 is the ratio of the risk of (non-epidemic) diarrhoeal illness in populations which have had a hygiene intervention compared to that in populations without such intervention. The fact that it is below 1 means that hygiene interventions reduce the risk of diarrhoeal illness – in fact by 37 per cent in this case ((1–0.63)×100). The confidence interval does not cross 1, hence providing evidence of reduction (as opposed to the play of chance).

2 Across the range of interventions there is overlap between confidence intervals, and therefore there is no clear evidence that one form of intervention is more beneficial than others.

3 The authors noted that 'the effect of multiple interventions does not seem to be additive' – meaning that the effect of two interventions applied together is lower than one might predict based on their individual effects: they do not add up. Given also the strong overlap in the effect estimates for the different forms of intervention, the choice of the most appropriate intervention is likely to be context-specific, and will depend on a range of other factors, including cost, acceptability and practicability.

The global perspective

To conclude this chapter, read a summary of facts and figures about water, sanitation and health provided by the WHO (2004). Issues of water and sanitation remain among the major challenges to health worldwide, and this is reflected in targets for the Millennium Development Goals. The problems have long been recognized, but to date our progress in tackling them has been disappointing. From a research perspective, the primary need is not so much to understand the epidemiological links as to identify ways to ensure affordable effective solutions are implemented in all communities without these basic resources.

 Water, sanitation and health

Water and Sanitation is one of the primary drivers of public health. I often refer to it as 'Health 101', which means that once we can secure access to clean water and to adequate sanitation facilities for all people, irrespective of the difference in their living conditions, a huge battle against all kinds of diseases will be won.

Dr LEE Jong-wook, Director-General, WHO

PROBLEMS

Diarrhoea

- 1.8 million people die every year from diarrhoeal diseases (including cholera); 90% are children under 5, mostly in developing countries.

- 88% of diarrhoeal disease is attributed to unsafe water supply, inadequate sanitation and hygiene.
- Improved water supply reduces diarrhoea morbidity by between 6% to 25%, if severe outcomes are included.
- Improved sanitation reduces diarrhoea morbidity by 32%.
- Hygiene interventions including hygiene education and promotion of hand washing can lead to a reduction of diarrhoeal cases by up to 45%.
- Improvements in drinking-water quality through household water treatment, such as chlorination at point of use, can lead to a reduction of diarrhoea episodes by between 35% and 39%.

Malaria

- 1.3 million people die of malaria each year, 90% of whom are children under 5.
- There are 396 million episodes of malaria every year, most of the disease burden is in Africa south of the Sahara.
- Intensified irrigation, dams and other water related projects contribute importantly to this disease burden.
- Better management of water resources reduces transmission of malaria and other vector-borne diseases.

Trachoma

- 500 million people are at risk from trachoma.
- 146 million are threatened by blindness.
- 6 million people are visually impaired by trachoma.
- The disease is strongly related to lack of face washing, often due to absence of nearby sources of safe water.
- Improving access to safe water sources and better hygiene practices can reduce trachoma morbidity by 27%.

Intestinal helminths (ascariasis, trichuriasis, hookworm)

- 133 million people suffer from high intensity Intestinal helminths infections, which often leads to severe consequences such as cognitive impairment, massive dysentery, or anaemia.
- These diseases cause around 9400 deaths every year.
- Access to safe water and sanitation facilities and better hygiene practice can reduce morbidity from ascariasis by 29% and hookworm by 4%.

Japanese encephalitis

- 20% of clinical cases of Japanese encephalitis die, and 35% suffer permanent brain damage.
- Improved management for irrigation of water resources reduces transmission of disease, in South, South East, and East Asia.

Hepatitis A

- There are 1.5 million cases of clinical hepatitis A every year.

Arsenic

- In Bangladesh, between 28 and 35 million people consume drinking-water with elevated levels of arsenic in their drinking-water.

- The number of cases of skin lesions related to drinking-water in Bangladesh is estimated at 1.5 million.
- Arsenic contamination of ground water has been found in many countries, including Argentina, Bangladesh, Chile, China, India, Mexico, Thailand and the United States.
- The key to prevention is reducing consumption of drinking-water with elevated levels of arsenic, by identifying alternative low arsenic water sources or by using arsenic removal systems.

Fluorosis

- Over 26 million people in China suffer from dental fluorosis due to elevated fluoride in their drinking-water.
- In China, over 1 million cases of skeletal fluorosis are thought to be attributable to drinking-water.
- The principal mitigation strategies include exploitation of deep-seated water, use of river water, reservoir construction and defluoridation.

DRIVING FORCES

Access to water supply as of 2002

- In 2002, 1.1 billion people lacked access to improved water sources, which represented 17% of the global population.
- Over half of the world's population has access to improved water through household connections or yard tap.
- Of the 1.1 billion without improved water sources, nearly two thirds live in Asia.
- In sub-Saharan Africa, 42% of the population is still without improved water.
- In order to meet the water supply MDG target, an additional 260 000 people per day up to 2015 should gain access to improved water sources.
- Between 2002 and 2015, the world's population is expected to increase every year by 74.8 million people.

Access to sanitation as of 2002

- In 2002, 2.6 billion people lacked access to improved sanitation, which represented 42% of the world's population.
- Over half of those without improved sanitation – nearly 1.5 billion people – live in China and India.
- In sub-Saharan Africa sanitation coverage is a mere 36%.
- Only 31% of the rural inhabitants in developing countries have access to improved sanitation, as opposed to 73% of urban dwellers.
- In order to meet the sanitation MDG target, an additional 370 000 people per day up to 2015 should gain access to improved sanitation.

Emergencies and disasters

- Almost 2 billion people were affected by natural disasters in the last decade of the 20th century, 86% of them by floods and droughts.
- Flooding increases the ever-present health threat from contamination of drinking-water systems from inadequate sanitation, with industrial waste and by refuse dumps.
- Droughts cause the most ill-health and death because they often trigger and exacerbate malnutrition and famine, and deny access to adequate water supplies.
- Disaster management requires a continuous chain of activities that includes prevention, preparedness, emergency response, relief and recovery.

Water resources development

- The development of water resources continues in an accelerated pace to meet the food, fibre and energy needs of a world population of 8 billion by 2025.
- Lack of capacity for health impact assessment transfers hidden costs to the health sector and increases the disease burden on local communities.
- Environmental management approaches for health need to be incorporated into strategies for integrated water resources management.

THE GLOBAL RESPONSE

Millennium Development Goals (MDGs)

By including water supply, sanitation and hygiene in the MDGs, the world community has acknowledged the importance of their promotion as development interventions and has set a series of goals and targets.

Goal 7: Ensure environmental sustainability

- Target 9: Integrate the principles of sustainable development into country policies and programs and reverse the loss of environmental resources.
- Target 10:
  - Halve by 2015, the proportion of people without sustainable access to safe drinking water and basic sanitation.
  - Integrate sanitation into water resources management strategies.
- Target 11: Have achieved by 2020, a significant improvement in the lives of at least 100 million slum dwellers.

Goal 4: Reduce child mortality

- Target 5: Reduce by two-thirds, between 1990 and 2015, the under-five mortality rate.

Goal 6: Combat HIV/AIDS, malaria and other diseases

- Target 8: Have halted by 2015 and begun to reverse the incidence of malaria and other major diseases.

Water for Life Decade: 2005–2015

- UN Declares 2005–2015 'Water for Life' as the International Decade for Action and set's the world agenda on a greater focus on water-related issues.

Salient quotations

We shall not finally defeat AIDS, tuberculosis, malaria, or any of the other infectious diseases that plague the developing world until we have also won the battle for safe drinking water, sanitation and basic health care.

Kofi Annan, United Nations Secretary-General

The human right to water entitles everyone to sufficient, safe, acceptable, physically accessible and affordable water for personal and domestic uses.

General Comment No. 15 (2002): The Right to Water

Summary

Water is a widespread and renewable resource, but increasing demand for it and pollution of water courses is leading to a potential global crisis of water management that may have implications for health. Approximately 40 per cent of the world's population face some level of water shortage. The availability of freshwater relative to water withdrawals can be an important constraint on development. Unless changes are made, it is estimated that, within the next two decades, the use of water by humans will increase by about 40 per cent and demand will outstrip available supplies. The growing problem of water scarcity adds to problems of inadequate access to clean water for much of the world's population. There are estimated to be 1.1 billion people who do not have access to an improved water supply and 2.4 billion who do not have adequate sanitation. A lack of access to clean water remains one of the greatest threats to global health, and arguably the largest of the environmental risks to health.

References

Cairncross S, Feachem R (1993). *Environmental Health Engineering in the Tropics: An introductory text*, 2nd Edition. Chichester, John Wiley & Sons.

Fewtrell L, Kaufmann RB *et al.* (2005). Water, sanitation, and hygiene interventions to reduce diarrhoea in less developed countries: a systematic review and meta-analysis. *Lancet Infectious Diseases* 5(1): 42–52.

Mehta L (2000). Water for the twenty-first century: challenges and misconceptions. IDS Working Paper 111 (available on-line at www.ids.ac.uk/).

Prüss A, Kay D *et al.* (2002). Estimating the burden of disease from water, sanitation, and hygiene at a global level. *Environmental Health Perspective* 110(5): 537–42.

UN/WWAP (United Nations/World Water Assessment Programme) (2003). *UN World Water Development Report: Water for People, Water for Life*. Paris, New York and Oxford, UNESCO (United Nations Educational, Scientific and Cultural Organization) and Berghahn Books.

UN-HABITAT (United Nations Human Settlements Programme) (2003). *Water and Sanitation in the World's Cities: Local Action for Global Goals*. London, Earthscan Publications Ltd.

UNDP (2002). *Global Environmental Outlook 3: Past, Present and Future Perspectives*. London, Earthscan Publications Ltd.

WHO (2004a). *Changing history, World Health Report 2004*. Geneva, World Health Organization.

WHO (2004b). *Water, Sanitation and Health*. Geneva, WHO.

Useful websites

UNESCO Water Portal: www.unesco.org/water/ (Provides links to the current UNESCO and UNESCO-led programmes on freshwater and is intended to serve as an interactive point for sharing, browsing and searching websites of water-related organizations, government bodies and NGOs, including a range of categories such as water links, water events, learning modules and other online resources)

World Health Organization: www.who.int/topics/water/en/ (Covers many health topics and the water webpage provides links to descriptions of activities, reports, news and events, as well as contacts and cooperating partners in the various WHO programmes and offices working on this topic. The site also contains links to related websites and topics)

10 | Water and health: wastewater use in agriculture

Overview

Municipal and industrial wastewater has long been used in agriculture and aquaculture, and extension of this practice may help relieve pressure on water supplies. In China, wastewater has been used to irrigate and fertilize crops for at least 3000 years, and in Vietnam it has been used in aquaculture for several centuries. The amount of wastewater used in irrigation is a matter of dispute and some estimates suggest that globally some 20 million hectares in 50 countries are irrigated with raw or partially diluted wastewater.

In this chapter you will consider the issues associated with using wastewater for irrigation of agricultural production, and the epidemiological evidence on whether such use leads to increased enteric disease. The chapter includes a practical exercise in which you will consider the measurement of risks to consumers of eating crops irrigated with treated wastewater – selecting the type of study and the scheme for exposure assessment.

Learning objectives

By the end of the chapter you should be able to:

- identify the principal issues associated with wastewater and excreta reuse, particularly in low-income countries
- develop an appropriate study design for an epidemiological study to assess the risk to consumers of eating crops irrigated with treated wastewater

Key terms

Aerobic Living or taking place in the presence of air or oxygen.

Anaerobic Living or taking place without air or oxygen.

Biogas Gas consisting mainly of methane produced by anaerobic digestion of organic waste.

Coliforms A group of bacteria, some of which (faecal coliforms), are normally found in human and animal faeces.

Effluent Outflowing liquid.

Infective dose The number of pathogens which must simultaneously enter the body, on average, to cause infection.

Sewage Human excreta (faeces and urine) and wastewater, flushed along a sewer pipe.

Sullage Domestic dirty water not containing excreta, also called grey water.

 Activity 10.1

To improve your understanding of wastewater and facilitate more accurate estimates of usage, it is necessary to have some common agreement on what is meant by wastewater. Consider how you might classify the use of wastewater. Spend a few minutes to write down a short list of broad headings. At the same time list why treatment of wastewater might be needed before it can be used.

 Feedback

Water is used in a wide range of activities, many of which you noted in Chapter 9. When we use water for one activity (e.g. drinking water, flushing toilets or generation of electricity) there is always a certain amount of 'waste' water that needs to be disposed of. For many activities this wastewater is simply disposed of via the drainage system, and is usually discharged directly into a river system. However, certain wastewater cannot be disposed directly into river systems, as these waters will contain contaminates such as faecal matter (e.g. domestic toilet waste), and heavy metals and other chemicals (e.g. from various industrial activities). These wastewaters need to be treated and should be transferred to a suitable water treatment works. After treatment, this water may be redistributed for human consumption via the water supply system, used for other purposes (such as irrigation) or discharged into river systems.

In terms of a typology for wastewater, the following three broad categories have been proposed by van der Hoek (2004):

1 *Direct use of untreated wastewater:* application to land of wastewater directly from a sewerage system or other purpose-built wastewater conveyance system.

2 *Direct use of treated wastewater:* the use of treated wastewater where control exists over the conveyance of the wastewater from the point of discharge from a treatment works to a controlled area where it is used for irrigation.

3 *Indirect use of wastewater:* the planned application to the land of wastewater from a receiving water body. Municipal and industrial wastewater is discharged without treatment or monitoring into the watercourse draining an urban area. Irrigation water is then drawn from rivers or other natural water bodies that receive wastewater flows.

Mara (1978) identified the main aims of wastewater treatment:

1 The destruction of the causative agents of water-related diseases (e.g. Table 9.1). This is particularly important in areas where the major cause of morbidity and mortality is the improper disposal of human faeces.

2 To convert the wastes into a readily reusable resource and so conserve both water and nutrients.

3 To prevent the pollution of any body of water (groundwater or surface water) to which the effluent escapes after reuse, or into which it is discharged without reuse. The organic pollution of waters is especially undesirable as it interferes with (or may even prevent) the use of the water for drinking and other domestic, industrial or agricultural purposes; it interferes with aquatic life (notably fish); and may drastically disrupt the ecology of the surrounding area (especially in arid zones).

Further detailed discussion on how water is treated is beyond the scope of this chapter. However, it is important to bear in mind that there tend to be great differences between high-income and low-income countries. In many low-income countries there is not only an inadequate infrastructure to provide potable water and sanitation facilities for millions of people, but the infrastructure to deal with the wastes (and especially wastewater) produced by these communities is also lacking. In many of these communities, where there is an already high burden of water-related disease, this lack of infrastructure to treat wastewater is likely to add to this underlying burden of disease.

Use of wastewater for irrigation

Wastewater is seen as providing 'a low-cost alternative to conventional irrigation water; it supports livelihoods and generates considerable value in urban and peri-urban agriculture despite the health and environmental risks associated with this practice' (Scott *et al.* 2004).

Regulation of wastewater is important at the local and national level, but there is also an international dimension, as the

> trade in agricultural products across regions is growing and products grown with contaminated water may cause health effects at both the local and transboundary levels. Exports of contaminated fresh produce from different geographical regions can facilitate the spread of both known pathogens and strains with new virulence characteristics into areas where such pathogens are not normally found or have been absent for many years.
>
> (Carr *et al.* 2004)

The health risks associated with irrigation using wastewater are shown in Table 10.1.

Examples of how wastewater is used

Pakistan

In almost all towns in Pakistan that have a sewerage system, the wastewater is directly used for irrigation (van der Hoek 2004). A negligible proportion of this wastewater is treated and no clear regulations exist on crops that can be irrigated with wastewater. Vegetables are the most commonly irrigated crops, because they

Table 10.1 Summary of health risks associated with the use of wastewater in irrigation

Health threats			
Group exposed	Nematode infection	Bacteria/viruses	Protozoa
Consumers	Significant risks of *Ascaris* infection for both adults and children with untreated wastewater; no excess risk when wastewater treated to <1 nematode egg/l except where conditions favour survival of eggs	Cholera, typhoid and shigellosis outbreaks reported from use of untreated wastewater; seropositive responses for *Helicobacter pylori* (untreated); increase in non-specific diarrhoea when water quality exceeds 10^4 FC/100 ml	Evidence of parasitic protozoa found on wastewater-irrigated vegetable surfaces but no direct evidence of disease transmission
Farm workers and their families	Significant risks of *Ascaris* infection for both adults and children in contact with untreated wastewater; risks remain, especially for children when wastewater treated to <1 nematode egg/l; increased risk of hookworm infection to workers	Increased risk of diarrhoeal disease in young children with wastewater contact if water quality 10^4 FC/100 ml; elevated risk of *Salmonella* infection in children exposed to untreated wastewater; elevated seroresponse to Norovirus in adults exposed to partially treated wastewater	Risk of *Giardia intestinalis* infection was insignificant for contact with both untreated and treated wastewater. Increased risk of amoebiasis observed from contact with untreated wastewater
Nearby communities	*Ascaris* transmission not studied for sprinkler irrigation but same as above for flood or furrow irrigation with heavy contact	Sprinkler irrigation with poor quality water 10^4 FC/100 ml, and high aerosol exposure associated with increased rates of viral infection; use of partially treated water 10^4 FC/100 ml or less in sprinkler irrigation not associated with increased viral infection rates	No data for transmission of protozoan infections during sprinkler irrigation with wastewater

Source: Carr *et al.* (2004)

fetch high prices in the nearby urban markets. The wastewater used for irrigation is valued by farmers mainly because of its reliability of supply. In some cases the wastewater is auctioned by the municipal council to the highest bidder, often a group of richer farmers who then rent out their fields to poor landless farmers. Under these conditions, the use of untreated wastewater is considered a win-win situation by both the authorities that are responsible for wastewater disposal and the farmers who get a reliable supply of water with high nutrient content. There are therefore very few incentives to invest scarce resources in wastewater treatment.

Mexico

Mexico accounts for about half of the 500,000 hectares irrigated with wastewater in Latin America. Much of the recent scientific work on health impacts and other aspects of wastewater use has been done in Mexico. In most cases the wastewater is used at some distance from the urban centre in a formal irrigation setting. The bulk of the untreated wastewater from Mexico City goes to Mezquital, immediately north of the Mexico Valley where it is used for irrigation via an extensive network of irrigation canals. This is probably the largest and longest-standing wastewater use system in the world.

Epidemiological evidence

A number of studies in Israel in the 1970s and 1980s focused on the health impacts of wastewater reuse in irrigation for agricultural purposes (Katzenelson *et al.* 1976; Fattal *et al.* 1986; Shuval *et al.* 1989). These studies were conducted in various kibbutzim (voluntary collective communities), and considered the health impacts of the aerosolized transmission of pathogens disseminated by wastewater sprinkler irrigation systems. The studies did not consider the health impacts that might be associated with the consumption of foodstuffs produced by these irrigation systems.

Katzenelson *et al.* (1976) based their study on an analysis of official reported cases of communicable disease taken from Ministry of Health records, and their work suggested increases in various enteric diseases including salmonellosis, shigellosis, typhoid fever, and infectious hepatitis. A second study did not confirm these findings, but did find a significant excess of enteric diseases, mostly in the 0–4-year age group (Fattal *et al.* 1986). When year-round rates for enteric disease were compared with those for the summer irrigation period (when most kibbutzim irrigate), and between exposed and control communities, these rates were found to be 'essentially the same' (Fattal *et al.* 1986).

The third study (Shuval *et al.* 1989) consisted of 20 kibbutzim, which were divided into three categories: (a) higher exposure (seven kibbutzim) – wastewater sprinkler irrigation within 300–600 metres of residential areas; (b) lower exposure (six kibbutzim) – with wastewater use, but no exposure to aerosols; and (c) no exposure (seven kibbutzim) – control group, with no use of wastewater. The results of the study are shown in Table 10.2.

When the year-round rates in the 'high exposure' group were compared with the 'no exposure' group there was virtually no difference, and this was also true when the data was disaggregated for the 0–5-year age group (see shaded areas in table). Similarly, when the rates during the irrigation period were compared there were no significant differences. Although the table suggests that there may have been a difference between the 'high exposure' and 'no exposure' group during the irrigation period (49.1/100 versus 36.6/100), this was not statistically significant (p-value 0.22). Thus, the authors concluded that 'the evidence provided by this study does not point to any significant health risk associated with exposure to treated wastewater aerosols generated at distance of some 300–600 m from residential areas'.

Table 10.2 Episode rate of enteric conditions (per 100 person-years) by kibbutz categories (all ages and 0–5, both sexes)

Age group	Category	(No.)	Population	Episodes Year* N	Rate	Irrigation period N	Rate**
All ages	Higher exposure	(7)	3,562	412	11.6	220	18.5
	Lower exposure	(6)	2,684	251	9.4	135	15.1
	No exposure	(7)	3,985	439	11.0	229	17.2
	Total		10,231	1,102	10.8	602	17.7
0–5	Higher exposure	(7)	538	140	26.0	88	49.1
	Lower exposure	(6)	436	87	20.0	52	35.1
	No exposure	(7)	656	173	26.4	80	36.6
	Total		1,630	400	24.5	220	40.5

* From March 1981 to February 1982
** Four months (May–August 1981) episodes $N \times 3 \times 100/p = $ rate/year
Source: Shuval et al. (1989)

✎ **Activity 10.2**

Your task is to design a study to assess the consumer risks from use of partially-treated wastewater.

The setting
The map (Figure 10.1) shows the area where the study is to be based. It shows the principal geographical features, the areas where crops are irrigated with water of different quality, and the locations of populations (villages, towns), markets and health centres. The water in the Tula River is primarily wastewater from a nearby large metropolitan area, which is not shown on this map.

As the Tula river flows northwards from the Requena reservoir towards the Endho Reservoir the quality of the water is 10^8 fc/100ml. The main branch of the river then takes a north-easterly course towards a group of communities that include Tezontepec and Progreso. At this stage the quality of the water has improved to 10^4 fc/100ml. As the Tula River reaches the main market town of Ixmiquilpan, there is a secondary branch (Teatote channel) where the quality of the water has improved to 10^3 fc/100ml. In the immediate vicinity water is used to irrigate vegetables.

To the west of Ixmiquilpan is the market town of Alfajayucan, which is surrounded by villages and irrigated land where vegetables are grown with partially-treated waste-water. This water comes via the Endho and Rojo Gomez Reservoirs, and by the time it reaches Alfajayucan the quality has improved to 10^3 fc/100ml. To the south-east is a group of villages where untreated wastewater is used, and to the north-west (surrounding the market town of Tecczaulta) is a group of villages where vegetables are irrigated with borehole water.

Your task
The objective is to design a study in this setting to address the following question: is there a risk of enteric infection from consuming crops eaten raw that are irrigated with partially untreated water?

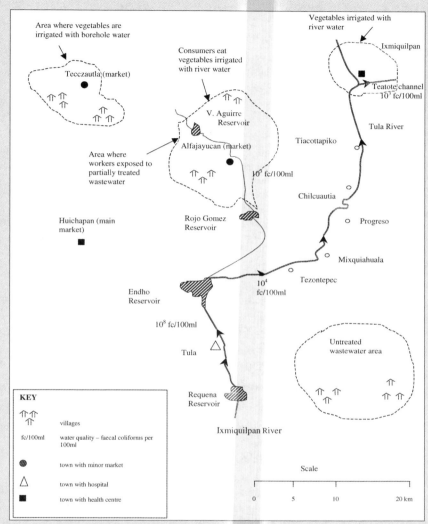

Figure 10.1 Sketch map of the La Cruz District
Source: Cifuentes et al. (2000)

In proposing your design, you should consider the following points:

- What type of study would you choose and why?
- What are the exposures you need to measure and how will you do so?
- What comparison group(s) would you choose?

Use whatever design you think most appropriate, but remember to think about practical considerations, and you should try to use a design that will achieve your objective as efficiently as possible. If you can, discuss your ideas with a colleague.

Feedback

Type of study design

Various possible study designs could be used to study the health effects of wastewater reuse, and they each have their advantages and disadvantages. However, probably the most efficient design in this situation would be a *cross-sectional survey*, which could compare the risks of enteric disease between those in the 'partially-treated' waste-water group, with those in the 'untreated wastewater' area, and those who use borehole water.

In many situations, cross-sectional studies are considered the least desirable of the standard epidemiological designs, but in this case a cross-sectional approach is probably the most efficient and practical. If you were to use a cohort design, you would have to carry out some form of (repeat) survey anyway in order to ascertain who develops illness. And the same would probably apply to a case-control study, unless you were to take people presenting with relevant illnesses to a health centre. But in this case, selection factors might make it difficult to have unbiased comparisons.

Because relevant illnesses (diarrhoea etc.) are common, you would pick up a reasonable number of cases on a single survey. And it is helpful that, for diarrhoea, the incubation period is usually short and duration of illness comparatively brief. Asking about when symptoms began would also enable you to distinguish newly-developed cases from chronic forms of diarrhoea.

Thus, with a cross-sectional survey, it would be logistically fairly straightforward to obtain evidence to relate current/recent diarrhoeal illness to recent dietary factors.

Exposure measurement

As indicated in Figure 10.1 there are three types of water usage in this region:

- untreated (e.g. villages in the south-east)
- partially-treated (e.g. Ixmiquilpan and Alfajayucan)
- borehole (e.g. Tecczaulta)

We are primarily interested in comparing the risk of diarrhoea among those who eat vegetables grown in areas where partially-treated wastewater is used with that of people consuming vegetables from other sources. However, there are several complicating factors. These include:

- what water sources are used to help *prepare* the food
- whether the food is cooked before eating (thereby removing or reducing risk)
- source of drinking water
- what type and quantity of vegetables are eaten

The number of pathogens ingested is likely to be influenced by:

- the amount of wastewater of varying microbial quality that is likely to cling to the surface of the various vegetables that would be consumed – this would probably be estimated from laboratory studies or other epidemiological studies
- the amount of pathogen that will die (e.g. UV irradiation from sunlight, washing vegetables in the home)

Vegetables with a large surface area and those that are difficult to clean, such as lettuces, may carry higher risk than those, like onions, which usually have their outer layers discarded. The survey questionnaire would therefore need to be carefully designed to elicit as much relevant information from the sample population as possible. It would need to establish, for example, where vegetables are bought; type of vegetables consumed; frequency and method of consumption; which individuals in the household eat this produce; whether there are seasonal variations in levels of consumption, and so forth. It might be helpful to construct, retrospectively, a sort of diary of foods and their sources for the last few days.

Comparisons

Although one might imagine comparing the risk of diarrhoea in populations from different areas (an 'ecological' analysis), this is less desirable than making measurements of exposure based on individual-level assessments. Even if an individual comes from an area where wastewater is used, the householder might obtain most vegetables from markets selling vegetables from other sources. The individualized approach also enables you to develop a more sophisticated semi-quantitative classification of exposure, based on sources, food types, quantities etc.

It would also be important to consider *potential confounding factors* such as:

* sources of drinking water
* occupation of workers in the family
* age
* socioeconomic and educational status

Data would need to be collected on these at the time of the fieldwork, and also incorporated into adjusted analyses.

Summary

You have learnt about definitions and types of wastewater, the use it is put to in irrigation and the potential health risks. You saw the levels of actual adverse effects that have been reported and went on to learn about how to undertake a study of assessing consumer risks.

References

Carr RM, Blumenthal UJ *et al.* (2004). Guidelines for the safe use of wastewater in agriculture: revisiting WHO guidelines. *Water Science Technology* 50(2): 31–8.

Cifuentes E, Gomez M *et al.* (2000). Risk factors for Giardia intestinalis infection in agricultural villages practicing wastewater irrigation in Mexico. *American Journal of Tropical Medicine and Hygiene* 62(3): 388–92.

Fattal B, Wax Y *et al.* (1986). Health risks associated with wastewater irrigation: an epidemiological study. *American Journal of Public Health* 76(8): 977–9.

Katzenelson E, Buium I *et al.* (1976). Risk of communicable disease infection associated with wastewater irrigation in agricultural settlements. *Science* 194: 944–6.

Mara D (1978). Wastewater treatment in hot climates, in Feachem R, McGarry M and Mara D.

Eds., *Water, Wastes and Health in Hot Climates*. Chichester, English Language Book Society and John Wiley.

Scott C, Faruqui N *et al.* (2004). Wastewater use in irrigated agriculture: Management challenges, in developing countries, in Scott C, Faruqui N and Raschid-Sally L, Eds., *Wastewater Use in Irrigated Agriculture: Coordinating the Livelihood and Environmental Realities*. CAB International in association with the International Water Management Institute, and International Development Research Centre: 1–10.

Shuval H (1978). Public health considerations in wastewater and excreta re-use for agriculture, in R. Feachem, *et al.* *Water, Wastes and Health in Hot Climates*. Chichester, English Language Book Society and John Wiley.

Shuval HI, Wax Y *et al.* (1989). Transmission of enteric disease associated with wastewater irrigation: a prospective epidemiological study. *American Journal of Public Health* **79**(7): 850–2.

UNESCO/WWAP (United Nations/World Water Assessment Programme) (2003). UN World Water Development Report: Water for People, Water for Life. Paris, New York and Oxford, UNESCO and Berghahn Books.

van der Hoek W (2004). A framework for a global assessment of the extent of wastewater irrigation: The need for a common wastewater typology, in Scott C *et al.*, Eds., *Wastewater use in Irrigated Agriculture: Coordinating the Livelihood and Environmental Realities*. CAB International in association with the International Water Management Institute, and International Development Research Centre: 11–24.

Useful websites

International Development Research Centre (Canada) http://web.idrc.ca/en/ev–1–201–1-DO_TOPIC.html

(In this chapter there have been several references to the book by Scott, CA *et al.* 2004 (Scott, Faruqui *et al.* 2004) which contains much useful information on the topic of wastewater in agriculture. It is freely available as a downloadable e-Book from this website. If you wish to read more about the kinds of issues addressed in this chapter then you will find this book an excellent resource.)

SECTION 5

Climate change

Climate change: principles

Overview

The world's climate has always changed but there is now strengthening evidence that for the first time it is changing as a result of human activity. The rate of change is predicted to be rapid. Global climate models suggest that average temperatures are likely to increase by 1.4 to 5.8 degrees Celsius by the end of the twenty-first century. Extremes of weather are predicted to become more common, and sea levels to rise. These changes may affect the health of human populations through direct and indirect mechanisms. Climate change is considered to be one of the key environmental threats of the coming century, but how to respond to it remains widely debated. Assessment of the health burdens and needed responses requires the combination of epidemiological analysis of weather-health relationships with climate modelling.

Learning objectives

By the end of this chapter you should be able to:

- **give a broad overview of the causes of climate change**
- **describe the potential health impacts that may be associated with it and estimate the population burdens**
- **outline the potential strategies for reducing those impacts**

Key terms

Adaptation Strategies, policies and measures undertaken now and in the future to reduce potential adverse impacts of climate change.

Climate The average state of the atmosphere and the underlying land or water in a specific region over a specific time scale. Should be distinguished from 'weather', which is the atmospheric conditions in a specific place at a specific time.

Climate change A statistically significant variation in either the mean state of the climate or in its measurable variability, persisting for an extended period (typically decades or longer).

Climate variability Variability in the mean state and other statistics (such as standard deviations, the occurrence of extremes etc.) of the climate on all temporal and spatial scales beyond that of individual weather events.

> **Climate change mitigation** An anthropogenic (human) intervention to *reduce the sources* or *enhance the sinks* of greenhouse gases.

Evidence of a changing climate

The world's climate has always changed as a result of natural cycles and cata-strophic events, although we still have an imperfect understanding of the various driving forces. Milutin Milanokovic (1875–1958) first proposed a link between past climatic patterns and fluctuation in insolation consequent to the eccentricity of the earth's orbit around the sun (periodicity of <100,000 years), fluctuation in the tilt of the earth relative to its plane of orbit (periodicity of ~42,000 years) and precession of its axis of rotation (periodicity of ~26,000 years). Less predicable variations occur in solar activity (reflected in the frequency of sunspots) which can increase the amount of solar radiation reaching the earth's surface, while particles generated from volcanic activity, for example, may reduce it. However, current debates about global warming centre on the effect of greenhouse gases – specifically carbon dioxide (CO_2) – generated by human activity through the burning of fossil fuels. There is accumulating evidence that the effect of anthropogenic (i.e. man-made) greenhouse gases will be critical to the earth's climate over the next hundred years or so.

Reliable measurements of surface temperatures, available for the last 200 years or so, indicate that global temperatures have been increasing over that time period (Figure 11.1). The most recent (2001) report of the UN's Intergovernmental Panel on Climate Change (IPCC 2001) estimates that the global average land and sea surface temperature has increased by 0.6 ±0.2 °C since the mid–nineteenth century,

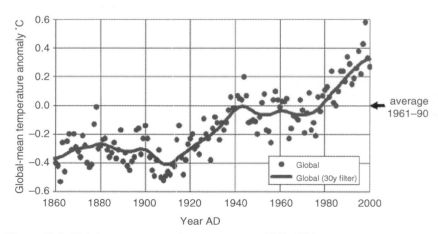

Figure 11.1 Global average near-surface temperatures, 1860–1999

Source: UK Hadley Centre

with most change occurring since 1976. The 1990s was the warmest decade on record.

Although this increase in temperature may be in part a natural trend following the Little Ice Age, current global climate models cannot account for the recent increase in temperatures without incorporating the effects of anthropogenic emissions of greenhouse gases. The IPCC has concluded that 'most of the warming observed over the last 50 years is likely to be attributable to human activities'. Moreover, based on a range of alternative scenarios, global climate models suggest that, if no specific actions are taken to reduce greenhouse gases, global temperatures are likely to continue to rise by between 1.4 to 5.8 °C by 2100 (compared to the 1990 baseline). Such a rise would be faster than any encountered since the inception of agriculture around 13,000 years ago. Predictions for precipitation and wind speed are less consistent, but also indicate significant changes.

To put these climate changes in context, it is worth examining the data about temperature changes over a geological time scale. Data from ice cores from the Vostok station in Antarctica (Petit *et al.* 1999) have been used to construct temperatures and carbon dioxide levels over the past 400,000 years or so (Figure 11.2).

There is a strong correlation between temperatures and atmospheric CO_2 levels, although the basis of the linkage remains unclear. We are now only around 6 °C warmer than during recent glacial periods, and CO_2 levels are already at their highest ever for this period. There is reasonable agreement among climate scientists that CO_2 levels are a critical driver for climate change over coming decades. Global emissions of CO_2 have risen exponentially over the last century and a half (Figure 11.3).

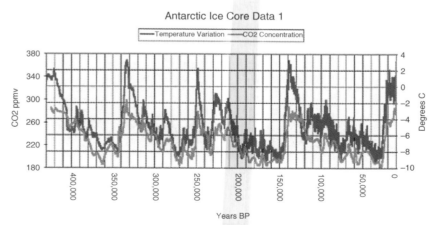

Figure 11.2 Climate and atmospheric history of the past 420,000 years from the Vostok ice core, Antarctica

Source: Petit *et al.* (1999)

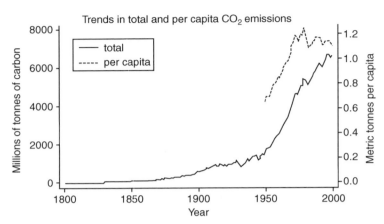

Figure 11.3 Trends in global CO₂ emissions

Source: Marland *et al.* (2005)

Importance of climate change for health

Figure 11.4 shows the range of predicted changes in global temperatures against the evidence of change in temperatures since the Ice Age. Estimates of the magnitude of the possible temperature change over the next hundred years are centred around 2 to 3 °C (but with a fairly broad range of uncertainty), which is substantial, but smaller than the change which has occurred since the Ice Age.

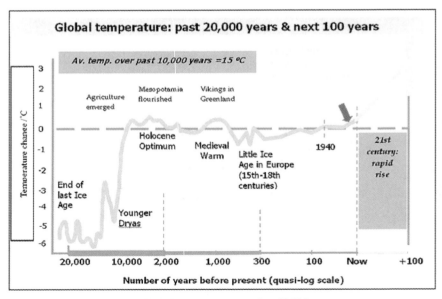

Figure 11.4 Change in mean global temperature over last 20,000 years

Source: McMichael AJ, Australia National University

Activity 11.1

Given that our climate is continually changing, why do you think the current predictions of climate change over this century are of concern, particularly in relation to health?

Feedback

There is uncertainty about the impacts of climate change for human populations, though we have a reasonable understanding of the sort of the changes that will occur, and the type of effects that may be linked to them. One of the debates is how important climate change will be, and whether we will be able to adapt to it. There are those who argue that humans are very resourceful, and will be able to find adequate technological solutions to climate change problems which are in any case unlikely to be severe. Others are much less optimistic, and are concerned that there will be substantial burdens, especially for populations in low-income countries who have the least capacity to adapt. There are several aspects of human-induced climate change that give rise to concern.

1 *The rapidity of change.* Although profound climatic changes have occurred over time, a rise of several °C over the next hundred years would represent very rapid change to which ecosystems would have little time to adapt (Figure 11.4). The temperature

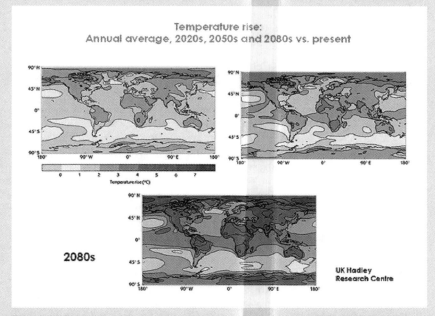

Figure 11.5 Models of change in global temperatures vs. present: 2020, 2050, 2080. These models suggest that there will be significant regional variation in climate change

Source: UK Hadley Centre

increase since the last Ice Age took thousands of years, and plants and animals could gradually migrate across latitudes, for example.

2 *Regional variations*. Although we refer to climate change as a single phenomenon, there are predicted to be significant regional variations in the change in temperature and precipitation, which may therefore produce more profound impacts for some populations (Figure 11.5).

3 *Vulnerability of fixed human settlements*. If there is significant climate change over this century, it will be the first period of major change since agriculture began. Modern societies are complex and sophisticated, but this may also make them more vulnerable. The location of cities is fixed, infrastructure can only be slowly changed and we have become more and more dependent on intensive methods of food production which might be disrupted in some areas.

 Activity 11.2

To understand the health consequence of climate change we have to be able to characterize and, if possible, quantify the relationship between climate and health.

1 What sorts of epidemiological studies will provide relevant evidence about climate change and health?

2 Why may they not give you the evidence that you really need to understand future health impacts of global warming?

 Feedback

1 The health impacts of climate change are difficult to quantify not only because they relate to an uncertain future, but also because there are complexities in defining the climate sensitivity of diseases. Epidemiological studies rarely, if ever, study the association between *climate* and health; most, instead, study *weather* and health. The distinction may seem semantic, but it has importance. The climate refers to the condition of a region with respect to prevailing meteorological conditions and is thus a time-averaged measure of temperature, precipitation, wind speed and the like; we think of warmer or colder climates, meaning ones where temperatures are generally warmer or colder. By weather, we mean something more immediate – meteorological conditions that occur at a particular point in space and time and which may change from day to day or even over a matter of minutes.

Although many study designs (cohort, case control etc.) have a role in epidemiological research into climate change, most studies are based on some form of time-series analysis in which short-term (usually daily or weekly) fluctuation in health is analysed in relation to fluctuation in temperature, rainfall etc. measured at similar temporal resolution. Such studies do not therefore relate health to prevailing climatic conditions, but rather to short-term changes in weather. They may establish that certain health parameters correlate with day-by-day changes in temperature, for example, but they do not say how health is affected over the medium term by living with an altered climate. There are strong parallels here with air pollution epidemiology (see Chapters 4 and 5). Just as the time-averaged effects of air pollution may not be the simple sum of the short-term associations determined by time-series studies, so too with climate and health. To study climate and health would mean comparing the long-term health experience of populations exposed to different (time-averaged) climatic conditions, but such comparisons would be difficult to interpret because populations differ for many reasons other than the climate they experience.

Time-series studies are used instead as an indirect measure of climate impacts. The first reason is that time-series studies are generally easier to perform and interpret. The same population is compared with itself day by day, so any change in health can be more reliably attributed to weather conditions if other time-varying confounders, such as air pollution, are taken into account. Second, most people would assume that if an excess of deaths occurs during a period of heat, for example, then that represents an important and potentially preventable health burden. Even though it reflects a short-term association between weather and health, it is taken as reasonable evidence of 'climate' impact as the frequency of (short-term) heatwaves is likely to increase under climate change.

2 The difficulty is that people may adapt to living with a warmer climate, such that they may not be as sensitive as expected to heatwaves and other weather events associated with the new climatic conditions. That adaptation may arise from physiological habituation, from alteration of behaviour or from structural changes to the built environment. Thus, the health effects of climate change may not be well represented by the results of studies of current short-term associations between weather and health – though they probably provide a reasonable indication. One of the epidemiological challenges in the field of climate change is to understand the degree to which adaptation is likely to reduce the impact of weather-health relationships.

Climate-related impacts

The types of impact that are expected to occur under climate change include direct effects (from thermal extremes, severe weather events, food and water-borne illness, changes in the distribution of vector-borne disease) and indirect effects (disruption of food production and water resources, social dislocation, reduced productivity). They have been described by a number of authors (McMichael et al. 2003) and are summarized in Figure 11.6. Understanding of the nature of the relationship of climate with health requires a range of study designs that will be described in the next two chapters.

The critical issues for epidemiological research are to understand how to adapt to climate change – how to limit its impacts on human health – and to provide the evidence to support arguments for reducing emissions – the strategy of 'mitigation'. The next two chapters focus on the epidemiological evidence that has a bearing on these issues.

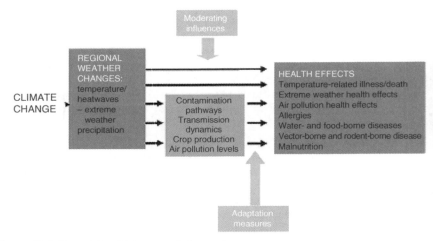

Figure 11.6 Potential health impacts of climate change
Source: after Patz *et al.* (2000)

Summary

There is now reasonable scientific consensus that the world's climate is rapidly changing as a result of man-made emissions of carbon dioxide. These changes may have health effects arising through the direct effects of heat and cold, the impacts of floods and storms, changes in the frequency of food- and water-borne disease and the altered distribution of vector-borne disease. Indirect impacts may arise from local and regional changes in weather, which affect crop productivity or economic activity. Assessment of the associated health burdens requires the combination of epidemiological analysis with climate modelling, but it entails many uncertainties. The nature of evidence is necessarily indirect for many, if not all, health impacts, as epidemiological studies do not investigate climate and health, less still climate change and health. Rather our understanding of vulnerability to climate change is based on studies of weather-health relationships whose importance in terms of climate change has to be judged from various indirect forms of evidence.

References

IPCC (2001). *Climate Change 2001: The Scientific Basis. The Contribution of Working Group 1 to the Third Assessment Report of the Intergovernmental Panel on Climate Change.* Houghton JT *et al.* New York, CUP.

Marland G, Boden T *et al.* (2005). Global, regional, and national CO2 emissions in *Trends: A Compendium of Data on Global Change.* Oak Ridge, Tennessee, Carbon Dioxide Information Analysis Center, Oak Ridge National Laboratory, U.S. Department of Energy.

McMichael A, Campbell-Lendrum D *et al.*, Eds. (2003). *Climate Change and Human Health: Risks and Responses.* Geneva, WHO.

Petit J, Jouzel J *et al.* (1999). Climate and atmospheric history of the past 420,000 years from the Vostok ice core, Antarctica. *Nature* **399**: 429–36.

Useful websites

The Hadley Centre for Climate Prediction and Research: www.metoffice.com/research/hadleycentre/
Intergovernmental Panel on Climate Change: www.ipcc.ch/
The Tyndall Centre for Climate Change Research: www.tyndall.ac.uk/

12 Climate change: extreme weather events

Overview

In this chapter you will look at the health effects of extreme weather events including heatwaves, floods and storms, which are predicted to increase in frequency under global warming. You will be asked to design a study to assess the impacts of flooding in an industrialized setting.

Learning objectives

By the end of this chapter you should be able to:

- describe the health impacts of extreme weather
- describe the ways of quantifying those impacts using time-series and other methods
- describe the concept and importance of effect modification with regard to climate change and its impacts on health

Key terms

Adaptive capacity The general ability of institutions, systems and individuals to adjust to potential damages, to take advantage of opportunities or to cope with the consequences of climate change in the future.

Extreme weather events Events that are rare within their statistical reference distribution at a particular place.

Scenario A description of a set of conditions, either now or, plausibly, in the future.

Vulnerability The degree to which individuals and systems are susceptible to or unable to cope with the adverse effects of climate change, including climate variability and extremes.

Climate change and extreme weather events

Extreme weather events (EWEs) are interactions between natural violence and society. The risk, i.e. the chance of something adverse happening, depends on both the weather hazard and vulnerability. Vulnerability is partly shaped by human decisions.

Many impacts of climate are related to EWEs and the same will hold for the impacts of climate change. Many studies of climate change impacts focus on changes in mean climatic conditions. However, global climate change is also likely to bring changes in climate variability and therefore more extreme events. The large damage potential of extreme events arises from their severity, suddenness and unpredictability, which makes them difficult to adapt to. Features of projected changes in extreme weather and climate events in the twenty-first century include more frequent heatwaves, less frequent cold spells (barring so-called singular events – see below), greater intensity of heavy rainfall events including increased flooding, more frequent mid-continental summer drought, greater intensity of tropical cyclones and more intense El Nino-Southern Oscillation (ENSO) events (IPCC 2001). Table 12.1 describes several extreme events that can substantially influence the vulnerability of sectors or regions to climate change.

Table 12.1 Typology of climate extremes

Type	Description	Examples of events	Typical method of characterization*
Simple extremes	Individual local weather variables exceeding critical level on a continuous scale	Heavy rainfall, high/ low temperature, high wind speed	Frequency/return period, sequence and/or duration of variable exceeding a critical level
Complex extremes	Severe weather associated with particular climatic phenomena, often requiring a critical combination of variables	Tropical cyclones, droughts, ice storms, ENSO-related events	Frequency/return period, magnitude, duration of variable(s) exceeding a critical level, severity of impacts
Unique or singular phenomena	A plausible future climatic state with potentially extreme large-scale or global outcomes	Collapse of major ice sheets, cessation of thermohaline circulation, major circulation changes	Probability of occurrence and magnitude of impact

* Stakeholders can also be engaged to define extreme circumstances via thresholds that mark a critical level of impact for the purposes of risk assessment. Such critical levels often are locally specific, so they may differ between regions.

Source: IPCC (2001)

Health impacts of extreme weather events

Extreme weather events brought about by global climate change are likely to have a wide range of health impacts. Many direct health impacts would result from changes in the frequencies and intensities of extremes of heat and cold and of floods and droughts. There would also be various health consequences of population displacement and economic disruption.

One difficulty in identifying impacts of EWEs is that the causation of most human health disorders is multifactorial and the 'background' socioeconomic, demographic and environmental context varies constantly. A further difficulty is foreseeing all of the likely types of future health effects, especially because for many of the anticipated future health impacts it may be inappropriate to extrapolate existing risk-function (exposure-response) estimates to climatic-environmental conditions not previously encountered. Estimation of future health impacts must also take account of differences in vulnerability between populations and within populations over time. Factors that affect vulnerability to disasters are shown in Figure 12.1.

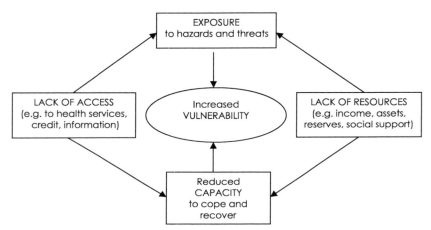

Figure 12.1 Diagrammatic illustration of vulnerability to disasters
Source: McMichael *et al.* (1996)

The increase in population vulnerability to extreme weather is primarily caused by the combination of population growth, poverty and environmental degradation (Alexander 1993). Concentration of people and property in high-risk areas (e.g. floodplains and coastal zones) also has increased. Degradation of the local environment may also contribute to vulnerability.

The health impacts of natural disasters include (Noji 1997; Ahern *et al.* 2005):

- physical injury;
- decreases in nutritional status, especially in children;
- increases in respiratory and diarrhoeal diseases resulting from crowding of survivors, often with limited shelter and access to potable water;
- impacts on mental health, which in some cases may be long-lasting;
- increased risk of water-related diseases as a result of disruption of water supply or sewage systems;
- release and dissemination of dangerous chemicals from storage sites and waste disposal sites into floodwaters.

Figure 12.2 shows the total number of deaths and of people affected by natural disasters by 100,000 inhabitants during the period 1974–2003 as recorded on the EM-DAT (2005) database. This illustrates that natural disasters, including those caused by extreme weather events, can directly result in many deaths and injuries, especially in economically disadvantaged populations. Furthermore, substantial indirect health impacts can also occur because of damage to the local infrastructure and population displacement. Following disasters, fatalities and injuries can occur as residents return to clean up damage and debris (Philen *et al.* 1992). Bereavement, property loss and social disruption may increase the risk of depression and mental health problems (WHO 1992).

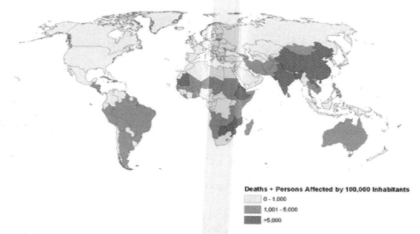

Figure 12.2 Total number of deaths and of people affected by natural disasters by 100,000 inhabitants, 1974–2003

Source: EM-DAT (2005)

Activity 12.1

Figure 12.3 shows the number of global natural disasters reported on the EM-DAT database.

1   The graph suggests an increase in the frequency of disasters over recent years, but what other reasons could the apparent increases be attributable to?
2   Certain years are marked with a black symbol; what irregularly-occurring natural phenomenon do these years represent?
3   Do you think there is enough evidence here to suggest that the frequency of global disasters increases during these specific years?

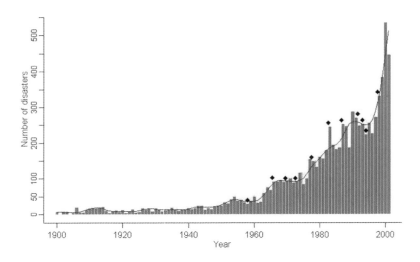

Figure 12.3 Number of reported global natural disasters
Source: EM-DAT (2005)

 Feedback

1 Changes in reporting onto this database over time may have caused some spurious increase. Experts have considered, however, that it is very likely that the frequency of natural disasters has increased in recent years, though the apparent rise may be accentuated (Kovats *et al.* 2003).

2 The black symbols represent El Nino years – El Nino is an irregularly occurring climate event that has wide-ranging consequences for weather around the world and is especially associated with droughts and floods.

3 By simple visual inspection of the graph, there does in general seem to be some evidence for an increase in the number of disasters recorded during El Nino years. This very informal analysis, however, is far from proof of an association.

Human health effects of flooding

A change in flood risk is frequently cited as one of the potential effects of climate change. It is clear that any increase in flooding will increase the risk of drowning, diarrhoeal and respiratory diseases, and in low-income countries, hunger and malnutrition.

On average about 50 million people are confronted with the consequences of flooding each year. Between 1971 and 1995, floods affected more than 1.5 billion people around the world. About 318,000 people died because of floods and more than 81 million were made homeless (The Dialogue on Water and Climate 2003). There were 26 'major flood disasters' worldwide in the 1990s, compared with 18 in the 1980s, 8 in the 1970s, 7 in the 1960s and 6 in the 1950s. The economic costs rose to an estimated US$300 billion in the 1990s, up from about $35 billion in the 1960s. Poor countries tend to suffer far more than wealthier ones when hit by weather disasters, both in terms of human casualties and economic loss.

 Activity 12.2

The health risks associated with flooding are surprisingly poorly characterized. Relatively little good epidemiological data is available on the effects of such events and so the full range of potential health impacts of flooding are uncertain.

Supposing you were asked to undertake a systematic analysis of flood-related health impacts, what kind of epidemiological study would you propose? Would you collect routine health data, or would you consider targeted interviews of affected people? How would you capture the potentially large mental health effects arising after a flood event? The specific aim of your study is to quantify the impacts of flooding on health and health services, and to provide the evidence base that will enable health and other authorities to improve their response to floods. A secondary aim is to quantify the extent to which flood-related health impacts are concentrated in vulnerable groups who might therefore gain particular benefit from targeted support.

 Feedback

The simplest approach would be to collect routinely-collected health data to determine whether increases in morbidity (and possibly also mortality) levels were elevated during and after a flood. The period of data collection should begin well before the actual event in order to establish baseline levels on health measures to compare with those during and after the flood.

In a high income country the biggest effects of a flood are likely to be on mental health. In order to fully identify these effects, it would be necessary to go beyond use of routine health data. Interviews of flooded victims would have to be conducted. This would require victims to be interviewed soon after the flood event, and ideally also at various time points post-flood as mental health effects of flooding can persist for months, or in some cases years, after the event took place. A suitable control group should also be interviewed in order to minimize the effects of recall bias. The interview study will suffer from the absence of a pre-flood baseline measurement, but because it is impossible to predict what area will be flooded, this appears impossible to circumvent.

Human health effects of temperature extremes

Global climate change is likely to be accompanied by an increase in the frequency and intensity of heatwaves, as well as warmer summers and milder winters. Figure 12.4 shows a time-series of daily mortality and daily mean temperature in London in 1976. As with all years in this population, there is a clear seasonal

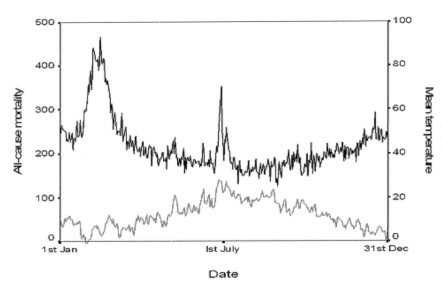

Figure 12.4 Time-series of daily mortality and daily mean temperature in London in 1976

pattern in both series, with death counts generally being at their highest in the winter months. The specific year of 1976 was unusual in that two large peaks in mortality occurred. The first peak was due to an influenza epidemic in the early part of the year, and the second associated with a 15-day heatwave in the middle of the year.

Deaths related to hot weather can occur not just during heatwaves. Figure 12.5 shows the relationship between daily mean temperature °C (x-axis) and the relative risk of all-cause mortality (y-axis) over a number of years in London. The relationship between temperature and mortality has been summarized using a smoothing spline (central estimate and 95 per cent confidence intervals). The risk of death increases at low temperatures (cold-related mortality) but also as temperatures rise. In the case of London, the heat-related deaths begin to occur once daily mean temperatures go above a threshold value of about 20°C. This type of U-shaped relationship is fairly typical in populations with a temperate climate, although the threshold value will likely vary between populations. However, increased mortality during periods of higher temperature has been observed in urban populations in tropical and sub-tropical settings, as well as in temperate climates.

The temperature measure shown in Figure 12.5 is an average value of levels on the day of death (lag 0) and the day before death (lag 1). This models the effect of temperature on mortality on the day of exposure and the day after. The effects of heat on mortality are known to be fairly immediate. Longer lags of temperature (two to three weeks duration) would be required to capture the full effects of cold temperature, the effects of which are known to be more delayed than for heat.

A more dramatic example of heat deaths occurred in the summer of 2003, when France and other countries of western Europe experienced exceptional temperatures which continued for over two weeks with comparatively little night-time

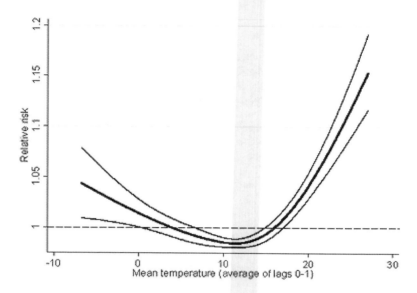

Figure 12.5 Temperature-mortality relationship in London

relief. With such exceptional temperatures (for France), deaths rose to several times their normal daily number, producing a very clear mortality peak once the heatwave had become established (Figure 12.6). In total, over 14,000 excess deaths occurred in France during this heatwave, and there were many other deaths in neighbouring countries which were also affected.

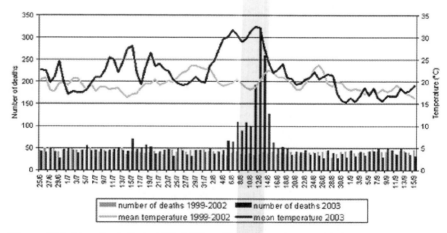

Figure 12.6 Mortality in Paris, 1999–2002 vs. 2003: effect of August 2003 heatwave

Source: Institut de Veille Sanitaire (2003)

✎ **Activity 12.3**

Unlike the assumptions made in the quantification of air pollution effects (see Chapter 4), the effects shown in the above relationships are not linear across the whole range of temperature. Instead a threshold value is usually identified, above which the temperature-mortality relationship is assumed to be linear (and similarly in reverse for cold temperatures). Table 12.2 shows results from a time-series study of temperature effects on mortality in Sao Paulo, Brazil. Shown are the relative risks (95 per cent confidence intervals) of cardio-respiratory mortality for every 1°C increase in temperature above 20°C (heat effect) and every 1°C decrease in temperature below 20°C (cold effect). The results are broken down by age group (children 0–14 years, adults 15–64, elderly 65+) and socioeconomic status.

Table 12.2 Relative Risks (RR) and corresponding 95% Confidence Intervals (CI) for cardiovascular (CVD) and respiratory (Resp) mortality, for heat and cold effects of mean temperature for a 1°C change above or below 20°C, by socioeconomic status and age group, Sao Paulo, Brazil, 1991–4

			Heat effect mean temperature (average lags 0–1) over 20°C			Cold effect weighted mean temperature (lags 0–20) below 20°C		
Cause	Age group	SES	RR	95% CI	Trend	RR	95% CI	Trend
CVD	adults	1	0.986	(0.971–1.002)		1.018	(1.006–1.031)	
		2	1.017	(0.999–1.035)		1.030	(1.016–1.045)	
		3	0.988	(0.960–1.015)		1.023	(1.002–1.045)	
		4	1.008	(0.973–1.044)	p=0.66	1.049	(1.022–1.077)	p=0.08
	elderly	1	1.018	(1.003–1.034)		1.059	(1.046–1.071)	
		2	1.031	(1.018–1.046)		1.069	(1.058–1.080)	
		3	1.015	(0.997–1.034)		1.005	(0.992–1.019)	
		4	1.006	(0.987–1.026)	p=0.36	1.066	(1.050–1.081)	p=0.78
Resp	adults	1	1.017	(0.989–1.046)		1.023	(1.000–1.046)	
		2	1.040	(1.007–1.074)		1.052	(1.026–1.079)	
		3	1.012	(0.961–1.065)		1.047	(1.008–1.088)	
		4	0.949	(0.885–1.018)	p=0.23	1.003	(0.954–1.055)	p=0.85
	elderly	1	1.026	(0.999–1.054)		1.055	(1.034–1.076)	
		2	1.014	(0.991–1.038)		1.052	(1.034–1.071)	
		3	1.033	(1.002–1.066)		1.084	(1.060–1.110)	
		4	1.031	(0.998–1.065)	p=0.64	1.067	(1.042–1.094)	p=0.21

Obs: SES=1 (most deprived) to SES=4 (most affluent)
Source: Gouveia et al. (2003)

1　Where do the biggest effects lie?
2　How do these relative risks compare with those from air pollution studies?
3　Which of the age groups seem most vulnerable in this population?
4　Is there a suggestion of effect modification by socioeconomic status? The study used district-level socioeconomic indicators rather than individual measures; how may this have affected the results?

5 The median value of temperature observed in the dataset was 19.6°C. Attributable risks were presented in the study; would these likely be bigger for the cold effects or the heat?

Feedback

1 The biggest relative risks were seen for cold effects in the elderly for both cardio-vascular and respiratory deaths. (In individual SES groups the heat effects were often not statistically significant, p>0.05.) However, the pattern is quite consistent, and in fact combining information across SES groups does show significant heat effects.

2 These relative risks appear of similar magnitude or slightly larger than those typically reported in time-series studies of outdoor air pollution. Remember, however, that they represent the change in risk for each degree Celsius change in temperature above the heat threshold, so the overall magnitude of effect associated with a heat episode is appreciable and larger than that associated with most air pollution episodes.

3 The results suggest that elderly people are more vulnerable than adults to both the effects of heat and cold.

4 There was little suggestion of effects varying by socioeconomic status – all statistical tests for trend (of RRs by SES level) were non-significant at the 5 per cent level. However, the use of area-level markers of socioeconomic status may have caused non-differential misclassification of SES, thus reducing an SES trend, if there is large socioeconomic variation within each area.

5 Since there were roughly as many days in this dataset above the threshold value of 20°C as there were below it, the higher relative risks associated with cold temperatures would mean larger attributable risks also.

Activity 12.4

Table 12.3 shows mortality statistics broken down by age and broad cause groups. The information is provided for three cities: New Delhi, London and Sao Paulo.

1 Which column is likely to correspond to which city?

Figure 12.7 shows the percentage change (95 per cent CI) in all-cause mortality

Table 12.3 Summary statistics of mortality breakdown in three cities (1991–4)

Percentage of deaths by age			
0–14	48.1	10.3	1.4
15–64	38.6	41.7	18.5
65+	13.3	47.7	80.0
Percentage of deaths by cause			
Cardiovascular	15.5	37.8	42.4
Respiratory	8.9	14.3	14.5
Other	75.7	47.9	43.1

Source: Hajat et al. (2005)

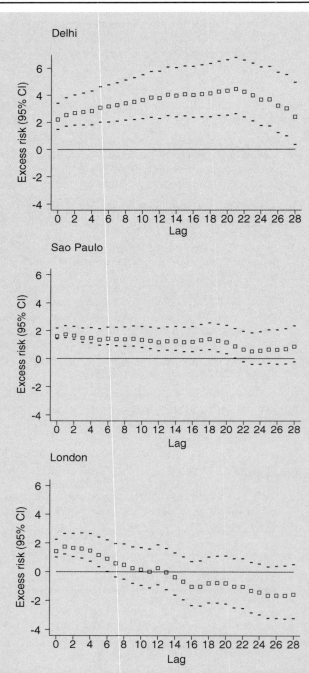

Figure 12.7 Excess risk (percentage increase in mortality per degree Celsius above the heat threshold of 20°C) summed over periods of increasing lag (unconstrained distributed lag models): all-cause mortality

Source: Hajat *et al.* (2005)

associated with high temperatures summed over periods of increasing lag. So, for example, the estimate at lag 0 represents the percentage change in same-day mortality associated with the temperature measure, the estimate at lag 1 represents the summed effect of temperature exposure on same-day mortality and also on mortality on the day after exposure etc. In the case of London, the heat effect 'peaks' at lag 1 after which time the summed effects reduce again. This suggests that days following high temperature days are associated with a reduced risk in mortality.

2   Why might that be the case? Why are different patterns observed in the other two cities, where the mortality breakdown by age and cause is so different?

 Feedback

1   Table 12.3 shows that there is wide contrast in the epidemiological profile of these three cities. The first column shows data for New Delhi where childhood deaths and deaths from infectious diseases are still very common. The third column represents London where the biggest proportion of deaths occur in the elderly and from cardio-respiratory causes. The profile for Sao Paulo (Column 2) was intermediate between New Delhi and London.

2   If heat-related deaths mainly occur in the elderly and in people already weakened by chronic diseases then these deaths may simply be being brought forward by a few days or weeks (mortality displacement or 'harvesting'). In such a situation, it would be reasonable to assume that an increase in mortality during high temperature days may be followed by a compensatory decline in the number of deaths a few days later – this concept was introduced for air pollution in Chapter 4. This would explain the reduction in the summed effect of temperature observed after lag 1 in the case of London in Figure 12.7. A similar decline was not apparent for New Delhi, where many heat-related deaths occur in children and in otherwise healthy individuals. Sao Paulo had an intermediate pattern of mortality displacement.

Even though high temperatures are associated with increased mortality in many populations, the same effects may not be apparent when considering morbidity outcomes. Figure 12.8 shows daily hospital admissions (seven-day moving average to remove large day of week variation) and daily deaths (all causes) in Greater London between 25 June 1995 to 1 September 1995. A heatwave occurred between 29 July–3 August 1995; during this time mortality increased but admissions remained largely unaffected. This suggests that many heatwave-related deaths occur in people before they come to medical attention. This has obvious implications for potential public health strategies.

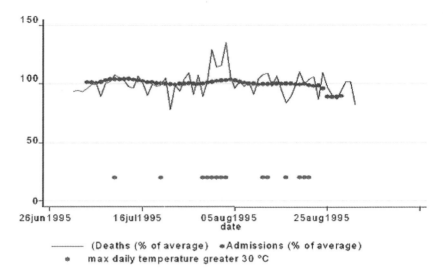

Figure 12.8 Daily hospital admissions (seven-day moving average) and daily deaths (all causes) in Greater London (29 June 1995 to 1 September 1995). Days with maximum temperature above 30°C are indicated

Source: Kovats *et al.* (2004)

Summary

You have looked at the health impacts of EWEs, which are predicted to rise in frequency and severity under climate change. Such events include storms and floods, but also episodes of high temperature which have been observed to increase the risk of mortality and other adverse health outcomes in most populations that have been studied to date. The frequency of EWEs giving rise to mortality and morbidity has risen over time, but largely reflecting reporting practice and changes in population vulnerability more than a climate change effect. The impacts of heatwaves can be studied by the same time-series methods that were discussed in relation to air pollution and health in Chapter 4, though the signal of increased deaths is sometimes clear from the standard time sequence if the rise in temperatures is exceptional. Epidemiological studies show that heat deaths occur after a very short time lag (usually same day or day after the heat), and while there is some evidence of mortality displacement for such deaths, this may depend on the population at risk and the nature of the heatwave.

References

Ahern M, Kovats R *et al.* (2005). Global health impact of floods: epidemiologic evidence. *Epidemiological Review* 27: 1–11.

Alexander D, Ed. (1993). *Natural Disasters*. London, University College London Press.

EM-DAT (2005). *The OFDA/CRED International Disaster Database*. Louvain, Belgium, University of Louvain.

Gouveia N, Hajat S *et al.* (2003). Socio-economic differentials in the temperature-mortality relationship in Sao Paulo, Brazil. *International Journal of Epidemiology* **32**: 390–7.

Hajat S, Armstrong B *et al.* (2005). Mortality displacement in heat-related deaths. A comparison of Delhi, Sao Paulo and London. *Epidemiology* **16**(4): 1–8.

Institut de Veille Sanitaire (2003). *Health Impact of the Heat Wave in France in August 2003.* Paris, Institut de Veille Sanitaire.

IPCC (2001). *Climate Change 2001: The Scientific Basis. The Contribution of Working Group 1 to the Third Assessment Report of the Intergovernmental Panel on Climate Change.* Houghton JT *et al.* New York, CUP.

Kovats R, Bouma M *et al.* (2003). El Nino and health. *Lancet* **362**: 1481–9.

Kovats RS, Hajat S *et al.* (2004). Contrasting patterns of mortality and hospital admissions during hot weather and heat waves in Greater London, UK. *Occupational and Environmental Medicine* **61**(11): 893–8.

McMichael A, Haines A *et al.*, Eds. (1996). *Climate Change and Human Health. WHO/EHG/ 96.7. An Assessment Prepared by a Task Group on Behalf of the World Health Organization, the World Meteorological Organization, and the United Nations Environment Programme, Geneva, Switzerland, 297.* Geneva, World Health Organization.

Noji E (Ed.) (1997). *The Public Health Consequences of Disasters.* New York, OUP.

Philen R, Combs D *et al.* (1992). Hurricane Hugo-related deaths – South Carolina and Puerta Rico, 1989. *Disasters* **16**: 53–9.

The Dialogue on Water and Climate (2003). *Climate Changes the Water Rules: How Water Managers Can Cope with Today's Climate Variability and Tomorrow's Climate Change.*

WHO (1992). *Psychological Consequences of Disasters.* WHO/MNH/PSF 91.3.Rev 1. Geneva, Switzerland, World Health Organization.

Useful websites

Institut de Veille Sanitaire (for description of health impacts of August 2003 heat-wave in France): www.invs.sante.fr/recherche/index2.asp?txtQuery=heat+wave, report in Eurosurveillance www.eurosurveillance.org/ew/2004/040311.asp#7)

World Health Organization, Geneva: www.who.int/topics/climate/en/

European Centre for Environment and Health, Rome: www.euro.who.int/ecehrome (contains links to several reports on climate change and adaptation)

13 | Climate change: vector-borne diseases

Overview

Climate change may affect the transmission of infectious diseases that are transmitted via an insect or tick vector. The disease pathogen is transmitted in the bite of the blood-sucking vector. In the case of malaria, the vector is a female Anopheles mosquito. Vector organisms are cold blooded and therefore sensitive to the local weather conditions. There is a minimum temperature (threshold) for activity, below which it is too cold for the vector to feed. At higher temperatures, the pathogen develops more quickly inside the vector and therefore the chance of an infective bite increases. Many factors other than climate are important determinants for the distribution of vector-borne diseases. Diseases such as malaria have been eradicated from many countries where the climate is suitable for transmission. Assessing the impact of changes in climate in the future need to take into account the capacity of countries to control the disease.

Learning objectives

By the end of this chapter you should be able to:

- describe the importance of temperature and rainfall in the transmission of vector-borne disease, and their role as determinants of the distribution of vector-borne diseases
- understand the different approaches to mapping the distribution of malaria (biological models, statistical models)
- describe the main approaches used for quantifying the potential impact of global climate change on malaria using climate scenarios
- describe the strengths and weaknesses of such approaches to assess risks to health in the future

Key terms

Vector An organism, such as an insect, that transmits a pathogen from one host to another.

Vector-borne diseases Diseases that are transmitted between hosts by a vector organism such as a mosquito or tick (e.g., malaria, dengue fever, leishmaniasis).

Vectorial capacity The average number of potentially infective bites of all vectors feeding upon one host in one day, or, the number of new inoculations with a vector-borne disease transmitted by one vector species from one infective host in one day.

Vector-borne diseases: epidemiology

Protozoa, bacteria and viruses transmitted by biting insects are among the most important causes of ill-health in low- and middle-income countries (Table 13.1). Many important tropical diseases are transmitted by mosquitoes.

Malaria is the world's most important vector-borne disease. The World Health Organization (WHO) estimates that about 2400 million people are at risk of infection with malaria (approximately 40 per cent of the world's population). Malaria is currently endemic in over 90 countries (WHO 2004). Of all infectious diseases, malaria continues to be one of the biggest contributors to the global disease burden in terms of death and suffering because of the large population at risk of endemic malaria in sub-Saharan Africa.

Table 13.1 Vector-borne diseases sensitive to climate factors

Vector	Diseases
Mosquitoes	Malaria, filariasis, dengue fever, yellow fever, West Nile fever
Sandflies	Leishmaniasis
Triatomines	Chagas' disease
Ixodes ticks	Lyme disease, tick-borne encephalitis
Tsetse flies	African trypanosomiasis
Blackflies	Onchocerciasis

 Activity 13.1

To begin, think about how you might investigate the role of climate as a determinant of the distribution of vector-borne disease. What type of information (data) would you like to collect, and how would you use it?

 Feedback

You might consider using data from the laboratory or the field (epidemiological). Many studies have been done in the laboratory of the effects of temperature and humidity on disease vectors and pathogen biology. Mosquitoes can be kept in climate-controlled conditions and the effect of temperature on their survival and feeding activity can therefore be studied under experimental conditions. For malaria, this has provided us with useful information about the dependence of vector and the plasmodium parasite on climatic parameters.

Alternatively, information can be obtained from studies of the vector and malaria cases in the real world through monitoring systems and dedicated epidemiological studies. However, such data can be difficult to obtain and their quality must be carefully assessed if used for comparative analysis. If data are available from more than one area, then the spatial relationship with climate could, in theory, be assessed, but it would be important to obtain data from a multitude of areas of differing climates. Useful evidence might also be gained by looking at variation in disease incidence from year to year. Cases of malaria can be obtained from the clinics where people are treated. Data on climate and weather can be obtained from the National Weather Services.

 Activity 13.2

Figure 13.1 describes the relationship between temperature and transmission factors for mosquito vectors of malaria.

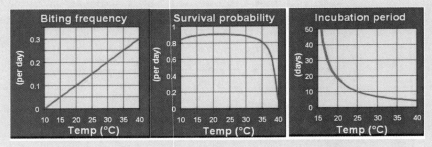

Figure 13.1 Temperature dependence of transmission factors for malaria parasite and mosquito vector

Source: Martens (1998)

- biting frequency = the number of blood meals (bites) made by the mosquito per day, on average; a biting frequency of 1 means that the mosquito feeds once every day
- survival probability = the probability of the mosquito surviving one day
- incubation period = the time it takes for the pathogen to develop in the mosquito before it can be transmitted in the next blood meal

1   What type of experiments do you think were used to generate these data?
2   What effect will increases in temperature have on the transmission of malaria?

 Feedback

1 These relationships are derived from data from experiments in the laboratory. They may not reflect the effect of ambient temperature in the field. Other factors, such as predation, will affect survival probability.

2 The incubation period of the parasite in the malaria mosquito (the extrinsic incubation period) must have elapsed before the infected vector can transmit the parasite. The parasites develop in the vector only within a certain temperature range. The minimum temperature for parasite development lies between 14.5°C and 15°C in the case of *P. vivax* and between 16°C and 19°C for *P. falciparum*. As temperatures increase, the time for parasite development decreases almost exponentially. The mosquito is more likely to transmit an infective bite before it dies. Biting frequency increases as the temperature increases because mosquitoes are cold blooded and their activity is related to the ambient temperature. Further, the mosquito needs to feed in order to avoid dehydration. As temperatures increase beyond the optimum range for the

mosquito, its chances of survival decreases, as it becomes too hot for the mosquito. Overall, these experiments indicate an optimum temperature range for the mosquito, and thus the potential for climate change to alter the transmission dynamics in some regions.

Climate models, climate change and malaria

Changes in climate would be expected to have the following effects on malaria:

- increase its distribution where it is currently limited by temperature – malaria may become present in new areas;
- decrease its distribution where it becomes too dry for mosquitoes to be sufficiently abundant for transmission;
- increase or decrease the months of transmission in areas with 'stable' malaria; some areas may change from unstable to stable malaria, and some may change from stable to unstable malaria;
- increase the risk of local outbreaks (that is, local transmission in areas where disease is eradicated but vectors are still present).

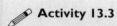 **Activity 13.3**

Read the following extract from Martens et al. (1999) and answer these two questions:

1 What information has been used to create this model?
2 What important information is missing from this assessment?

 Climate change and future populations at risk of malaria

Global estimates of the potential impact of climate change on malaria transmission were calculated based on future climate scenarios . . . This assessment uses an improved version of the MIASMA malaria model, which incorporates knowledge about the current distributions and characteristics of the main mosquito species of malaria. The greatest proportional changes in potential transmission are forecast to occur in temperate zones, in areas where vectors are present but it is currently too cold for transmission. Within the current vector distribution limits, only a limited expansion of areas suitable for malaria transmission is forecast, such areas include: central Asia, North America and northern Europe (Figure 13.2). Possible decreases in rainfall indicate some areas that currently experience year-round transmission may experience only seasonal transmission in the future.

Description of malaria model

Transmission potential (TP) and population at risk are the two main outcome measures of the malaria model. TP is a comparative index for estimating the impact of changes in environmental temperature and precipitation patterns on the risk of malaria. A high TP indicates that, despite a smaller vector population, or, alternatively, a less efficient vector population, a given degree of transmission may be maintained in a given area. TP is an estimate of the true vectorial capacity (VC) that changes from site to site, from vector to vector, and within and between transmission seasons. Absolute values of TP should be interpreted with caution as they are based on incomplete data concerning parameter values. TP is used to estimate the effect of a range of climate scenarios on malaria risk for three times (2020s, 2050s, and 2080s). All climate scenarios indicate an increase in global mean temperature but cooling does occur in some scenarios in small areas. The climate scenarios used in this assessment have considerable variability in the spatial distribution and magnitude of precipitation. Nearly all areas show an increase of malaria transmission potential as climate changes. A few areas indicate a decrease in transmission potential associated with decreases in precipitation.

The 'population at risk' is defined as the total population living in an area where conditions are suitable for malaria transmission, based on the climatological parameters. The reference scenario, from which additional population at risk estimates are calculated, is one that includes population growth but the climate is the baseline climatology 1961–90.

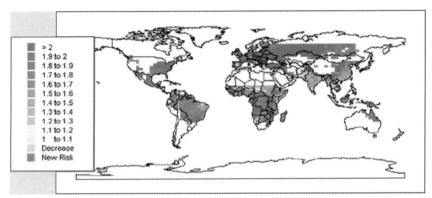

Figure 13.2 Change in transmission potential of falciparum malaria by 2080 under a climate change scenario (HadCM2GgaX)

Source: Martens et al. (1999)

⟳ **Feedback**

1 This paper illustrates one (early) approach that has been developed to assess the impact of climate change on malaria. 'Biological' or deterministic models of malaria are based on the relationships between temperature and the extrinsic incubation period of the parasite, and therefore the probability of completing the transmission cycle. These relationships are derived from laboratory data and are assumed to apply in the field. They are a valid approach as sensitivity analyses for relative changes in risk as temperature changes. However, these models are not ideal for describing the geographical distribution of malaria within endemic areas. Being based on laboratory data, the temperature relationships derived may not be appropriate to conditions in the field. The models assume that climate input data accurately represent the climatic conditions that mosquitoes and parasites experience in the field, disregarding the possibility that vectors might use microhabitats (e.g. shelter in houses or trees) that are very different from temperature measured at the meteorological station. The outputs from biological models should be validated against current disease distributions to provide useful information for assessment.

2 Information on factors other than climate that will affect the distribution of malaria is missing from this model. The most important factors are those that affect disease control in the future. It is very likely that malaria will decline due to improvements in treatment and control. *Adaptive capacity* is defined as a population's ability to cope with the impacts of climate change in the future. Thus, a population with a high adaptive capacity will be able to cope with (control) the additional risk of malaria due to climate change. Vector abundance is not included in this model although it is an important determinant of transmissions (and is also affected by climate). However, it is very difficult to measure at the national or regional scale, and global data are not available.

Rainfall and malaria

As well as temperature, rainfall has an important effect on disease transmission. Mosquitoes breed in standing water (usually freshwater pools or marshes) and, therefore, mosquito abundance is affected by rainfall when water has collected in puddles and pools. Rainfall also affects relative humidity and hence the longevity of the adult mosquito.

✏ Activity 13.4

Figure 13.3 shows seasonal rainfall and malaria cases in a highland region of Africa:

1 What do these curves tell us about the relationship between rainfall and malaria?
2 What other possible explanations are there for the increases in malaria?

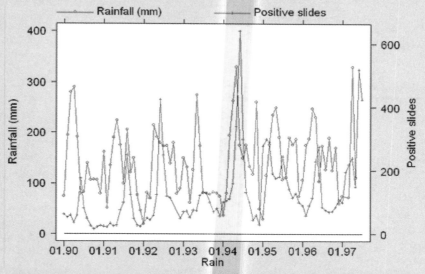

Figure 13.3 Relationship between rainfall and malaria cases (slides positive for the parasite) in a highland region 1990–7

Feedback

1 The temporal association is not easy to discern, but increases in rainfall appear to be associated with increases in malaria transmission after a time lag.

This pattern of malaria is described as 'epidemic' because transmission is greatly increased in short periods (epidemics) and then returns to a low level (though not to zero). In endemic malaria, transmission is more stable from year to year. Due to the importance of climate factors in transmission, malaria has some degree of seasonal pattern wherever it is found. We might guess that the observed delay between increases in rainfall and the reported increases in cases reflects some aspect of the biology of the vector and/or the parasite. In fact, we believe the delay is related to the breeding cycle: the mosquitoes lay their eggs in puddles formed after rainfall and it then takes time for the eggs to develop to adult blood feeding mosquitoes. The curve showing cases of malaria (slides of blood that contain parasites) is evidently based on data taken from a clinical setting – in fact a local clinic. The data will not therefore include all cases of malaria in the vicinity as not everybody with malaria will seek medical attention, but the week to week variation is probably a good guide to the temporal fluctuation in disease.

2 Other possible explanations for the malaria epidemics would need to be considered, and these include changes in vector control (insecticide spraying), or perhaps some migration into the area of people carrying the malaria parasite. It is important to note that malaria has unique features in each location, through its variety of vectors and the ecological conditions that favour transmission. It is useful to think of malaria as many diseases rather than one. Short-term unusual climate conditions in the opposite direction to those usually experienced (e.g. rainfall in arid regions and drought in more humid climates) may cause epidemics. It is therefore not possible to generalize the effects of rainfall on epidemic malaria, and caution is needed not to over-interpret apparent temporal associations seen in any one area based on comparatively short data series.

Mapping vector-borne diseases

Although at very broad scale there appears to be an association between malaria and climate, non-climatic environmental factors are likely to be equally, if not more, important in determining the distribution of disease. Such factors include land use (presence of possible breeding sites), level of socioeconomic development, degree of urbanization, quality of health care, public health infrastructure and whether control programmes are used.

A second method of examining the influence of meteorological and other environmental factors on disease distribution is to use disease mapping. This approach has been greatly enhanced by the advent of geographical information systems (GISs), and the use of (raster-based) satellite data on land cover and meteorological parameters which can be integrated with data gathered from the field on disease vectors (obtained using trapping methods etc.) and clinical cases. The principle is to assess the spatial or temporal-spatial correlation between the environmental (including meteorology) and disease (vector abundance, disease cases etc.) variables. The correlation may be analysed using various statistical techniques (Sutherst 1998; Rogers and Randolph 2000). If such evidence is then combined with models of climate change, quantitative predictions of the new geographical limits of vectors and disease under various climate scenarios can be obtained. Inevitably, such models entail many uncertainties, and they are sometimes criticized for this.

Although the statistical approach does not throw direct light on the mechanisms underlying climate sensitivity of vector-borne diseases, it generally provides plausible results because it suggests extension from the current observed distribution. However, the method heavily depends on the availability of high-quality geographical data on both vectors and disease.

 Activity 13.5

Read the following extract from Rogers and Randolph (2000) and then answer the questions below:

1 What are the different approaches between the Rogers and Randolph paper and the paper by Martens et al.?
2 What factors were not included in the model and hence contribute to uncertainty in the projections of future malaria impacts?

 The global spread of malaria in a future warmer world

The frequent warnings that global climate change will allow falciparum malaria to spread into northern latitudes including Europe and large parts of the United States are based on biological transmission models driven principally by temperature (Figure 13.4). These models were assessed for their value in predicting present and therefore future malaria distribution. In an alternative statistical approach the recorded present-day global distribution of falciparum malaria was used to establish the current multivariate climatic constraints. These results were applied to future climate scenarios to predict future distributions which showed remarkably few changes even under the most extreme scenarios.

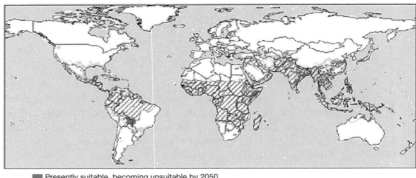

Presently suitable, becoming unsuitable by 2050
Presently unsuitable, becoming suitable by 2050

Figure 13.4 Net change in population at risk of falciparum malaria by 2050 under climate change (IS92a climate scenario)

Source: Rogers and Randolph (2000)

 Feedback

1 The global modelling study of Rogers and Randolph used a statistical approach based on the current distribution of malaria (which is, of course, determined by both climatic and non-climatic factors, including land use and level of socioeconomic development). Their map of future distribution, however, was based only on changes in temperature, rainfall and humidity. Using a particular climate scenario (known as IS92a), they estimated no significant net change by 2050 in the portion of world population living in actual malaria-transmission zones, specifically those suitable for *P. falciparum*.

2 The limitations of this technique centre on the fact that it does not allow separation of the influence of climatic and non-climatic factors. The Martens paper bases its models only on the climatic factors, and the map should therefore be interpreted cautiously, recognizing that non-climatic factors may be the more important determinant of the distribution of malaria.

Using scenarios to estimate future impacts

Scenarios provide an important tool for estimating the potential impact of climate change on specific health outcomes. Scenarios are not predictions of future worlds or of future climates. There are many ways of applying scenarios, which have been variously defined as:

* plausible and often simplified descriptions of how the future may develop, based on a coherent and internally consistent set of assumptions about driving forces and key relationships;
* hypothetical sequences of events constructed for the purpose of focusing attention on causal processes and decisions points;

- archetypal descriptions of alternative images of the future, created from mental maps or models that reflect different perspectives on past, present and future developments.

Climate scenarios are plausible representations of future climates that have been constructed for use in investigating the potential impacts of climate change. Many global climate scenarios are available that describe changes in climate at a spatial resolution of 0.5° grid. Climate scenarios must be linked to explicit emissions scenarios that project future greenhouse gas emissions (the main driver of climate change) (Figure 13.5).

A climate scenario is not the same as a climate projection. Climate projections (i.e. the results of experiments using a climate model driven by GHG emission scenarios) alone rarely provide sufficient information to estimate future impacts of climate change. Model outputs commonly have to be manipulated and combined with observed climate data to be useable within the vulnerabilities, impacts and adaptation research communities.

Scenarios can also be developed regarding possible changes in the adaptive capacity of the population of interest. In general, it is possible to identify a number of factors that determine adaptive capacity and to identify plausible states of those factors in the future. An obvious factor is income (and this often is measured at national level by GDP *per capita*). When assessing the impact of climate change, it is useful to consider two or three states for the future – reduced capacity as a result of deterioration in one or more of the determinants of adaptive capacity, similar capacity with little to no change in the determinants, and increased capacity as a result of an enhancement in one or more of the determinants. As in the case of all scenarios, the basis/assumptions for constructing these scenarios must be consistent and plausible in the light of the chosen future view (i.e. across the selected emission, climate, socioeconomic and adaptive capacity scenarios).

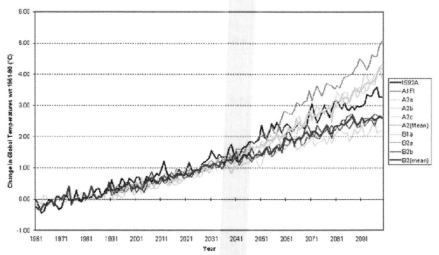

Figure 13.5 Global trends in near surface temperatures for four different emissions scenarios (IS92a, A1FI, A2, B1 and B2) and a range of climate model experiments

Source: UK Hadley Centre

Activity 13.6

What are the advantages and disadvantages of using scenarios in climate change impact assessment?

 Feedback

Scenarios are useful because they:

- describe key considerations and assumptions: scenarios can help to imagine a range of possible futures if a key set of assumptions and considerations is followed
- combine quantitative and qualitative knowledge: scenarios are powerful frameworks for using both data and model-produced output in combination with qualitative knowledge
- identify constraints and problems: exploring the future often yields indications for constraints in future developments and dilemmas for strategic choices to be made
- expand thinking beyond the conventional paradigm: exploring future possibilities that go beyond conventional thinking may result in surprising and innovative insights

But there are also weaknesses:

- lack of diversity: scenarios are often developed from a narrow, disciplinary-based perspective, resulting in a limited set of standard economic, technological and environmental assumptions
- extrapolations of current trends: many scenarios have a 'business-as-usual' character, assuming that current conditions will continue for decades
- inconsistency: the sets of assumptions made for different sectors, regions or issues are often not consistent with each other
- lack of transparency: key assumptions and underlying implicit judgements and preferences are not made explicit; for example, it may not be clear which factors or processes are exogenous or endogenous and to what extent societal processes are autonomous or influenced by concrete policies

Summary

You have considered the methods used to quantify the potential impact of climate change on vector-borne disease at the global scale. Most such modelling has focused on malaria and to a lesser extent on dengue fever and tick-borne diseases. Although we have reasonable knowledge of the underlying links between meteorology, vector and pathogen, and human infection, non-climatic factors have central importance as determinants of the distribution of disease in the real world. Because researchers have not yet been able to quantify their influence with precision, there is considerable debate over the extent to which changing patterns of temperature, rainfall and humidity will alter the distribution and occurrence of disease. A number of authors have pointed out that, if climate change has an appreciable effect on vector-borne disease, it is most likely to do so in areas of poor public health infrastructure on the margins of the current distribution of disease.

References

Martens WJ (1998). Health and climate change: modelling the impacts of global warming and ozone depletion. London: Earthscan.

Martens P, Kovats S *et al.* (1999). Climate change and future populations at risk from malaria. *Global Environmental Change* 9: S89–107.

Rogers D and Randolph S. (2000). The global spread of malaria in a future, warmer world. *Science* **289**: 1763–65.

Sutherst R (1998). Implications of global change and climate variability for vector-borne diseases: generic approaches to impact assessments. *International Journal of Parasitology* **28**: 935–45.

WHO (2004). *Water, Sanitation and Health*. Geneva, WHO.

Useful websites

Hadley Centre for information on climate change and climate models: www.meto.gov.uk/research/hadleycentre/models/modeltypes.html

National Oceanic and Atmospheric Administration (NOAA): www.noaa.gov/

WHO (malaria): www.who.int

SECTION 6

Epidemiological evidence

Reviewing epidemiological evidence

Overview

In this chapter you will consider the interpretation of scientific papers and reports on environmental hazards to health, which is based on the same principles as the interpretation of any epidemiological report. Time-series, geographical and quasi-experimental research designs are comparatively common in the environmental field, and common issues of interpretation are raised in relation to exposure measurement, control of confounding, small relative risks but large attributable burdens, and public health significance. A framework for interpretation is presented and illustrated using extracts for two papers.

Learning objectives

By the end of this chapter you should be able to:

- **summarize the key points of a paper or other research report on an environment and health issue**
- **discuss the main issues of design, conduct and analysis relevant to its interpretation**
- **make a broad assessment of its implications for public health**

A framework for interpreting papers and reports

We begin by presenting a framework – a checklist – for interpreting studies in the field of environmental public health. Most of the principles apply in all domains of epidemiological/public health research, but we emphasize some of the features that merit particular comment with environmental studies. The framework is meant as a guide only. You may wish to adapt it to suit your own purposes.

Summarizing the paper

As with any paper or report, a useful first step is to summarize the key elements of the paper. This summary could be divided into two parts:

- its title/subject, where was it published and who wrote it;
- its main design features and result, specifically:
    Design
    Setting
    Exposure

Outcome
Main result.

Although your assessment of its scientific quality should be based on the objective criteria, it can be useful to note the authorship and journal of publication. For one thing this may give a broad indication of how rigorously the paper has been peer reviewed, how widely its findings are disseminated and, to some degree, its scientific merit. It is wise to be cautious, however. Even prestigious journals are sometimes tempted to publish papers of comparatively low scientific rigour, especially if they address issues of current interest or controversy. Some of the most rigorous scientific papers appear in specialist journals where they are peer reviewed by experts in the relevant field.

It is useful to report the design features of the paper and its main result as doing so helps to crystallize its essential features and guide your further thoughts about potential strengths and weaknesses. Your assessment of the paper should lead you to be able to answer the following questions:

- Are the aims and objectives clearly stated?
- Are the methods suitable to answer the research question?
- Have exposure and outcome been appropriately measured?
- Are the conclusions justified based on the evidence?

Your answers to these will be informed by consideration of the headings listed below.

Is there association?

Having summarized the main features, the principal issue to consider is whether there is convincing evidence of association between the exposure and health outcome. In general terms this would mean thinking about bias, confounding and chance, each of which raises its own set of subsidiary questions.

Bias

The nature of the most critical form of bias depends on the study design. For example:

- cohort studies: losses to follow-up;
- case-control studies: selection of controls, recall of past events;
- cross-sectional studies: response rate;
- time-series studies: control for time-varying confounders;
- geographical studies: comparability of case ascertainment, exposure measurement, ecological bias.

Remember that avoiding bias is arguably the most important aspect of epidemiological studies as there is usually very little that can be done to correct or assess the influence of bias once it has occurred.

Confounding

In nearly all studies, it is useful to think about the 'common confounders' – age, sex, ethnicity, socioeconomic status, smoking, etc. – as well as more specific confounding factors that may be relevant to the outcome of interest. Have they been adequately dealt with in design (restriction, matching, stratification) or analysis (stratification/standardization, regression)? Is there potential residual confounding?

Chance

Assessing the role of chance can sometimes be very complicated, especially in studies which are not testing a single, clearly-stated hypothesis. Confidence intervals and inference tests are useful, but you should always bear in mind the context. Is the study hypothesis testing or generating? Was there a clear prior hypothesis or was this a *post hoc* analysis (recall the Texas sharp shooter phenomenon with cluster investigations)? The apparent statistical significance is often difficult to judge when there has been multiple testing or multiple subgroup analyses. Evidence of exposure response trends is often particularly useful. With negative studies it is appropriate to consider issues of power, though of course power cannot be calculated after the event.

Is the association causal?

The assessment of causality is really an extension of the issues of bias, chance and confounding. It is worth having in mind a set of criteria, such as those proposed by Bradford Hill in his classic paper (1965):

1 *Temporal sequence: did the exposure precede the outcome?* In environmental studies it is worth remembering that the time lag for the development of solid tumours may be as long as 10 to 20 years or more. Thus, if studying the health of a population living around an industrial plant, a raised incidence of lung cancer is very unlikely to be attributable to the plant if it has been operational for less than five years. This fact can sometime be turned to advantage by providing a 'control' period in analysis.

2 *Strength of association.* Strong associations (measured by ratios of rates, risks, odds) lend weight to a causal association as they are unlikely to arise from residual confounding. However, for many environmental exposures, relative risks are small. A good example is the relative risk determined from daily time-series studies of the association between outdoor air pollution and mortality. Typically, mortality is only a few per cent higher on the most polluted days compared with the least polluted – a relative risk of around 1.03. In most other contexts such a small relative risk would be considered almost uninterpretable. However, the specific design features of daily time-series mean that reasonably secure interpretation can be made of such small relative risks.

3 *Consistency of association.* Is the same association seen in other studies of similar and different design?

4 *Biological gradient.* Evidence of increasing disease risk with increasing exposure is often a quite persuasive (but by no means infallible) indicator of causal association, but it is worth noting that relationships can be non-linear.

5 *Specificity.* This refers to the fact that we normally expect a specific exposure to be associated with a specific health outcome(s), although there are examples of single factors (e.g. smoking) being associated with quite a broad range of diseases. You may recall that the specificity of impact of particle pollution

on cardio-respiratory rather than non-cardio-respiratory disease was one of the factors that strengthened the interpretation of the semi-ecological cohort studies of air pollution and health.

6 *Biological plausibility.* In practice, we often accept plausibility even in the absence of a mechanism unless the association seems highly counter-intuitive.

7 *Coherence.* Is the reported association compatible with the broader body of scientific knowledge?

8 *Experiment.* In most environmental studies, this relates to natural experiments where an existing exposure is removed, or a new exposure is introduced – usually assessed by quasi-experimental designs (interrupted times series and the like).

9 *Analogy.* Have other similar associations been demonstrated?

You may find that some of these criteria are more useful than others in assessing environment and health studies, and it is prudent not simply to add up the scores for each point, but rather to give most weight to criteria that appear to be most relevant in particular cases.

What is the public health importance?

Having decided on the overall strength of evidence, it is pertinent to consider the importance of the findings for public health. Some of the relevant headings here are:

- generalizability;
- clinical importance;
- population attributable risk;
- equity;
- scope for prevention/amelioration – cause and effect? reversibility;
- public perceptions.

Remember that generalizability of results does require the study to have been conducted on a 'representative' population. Inferences about biological effects are likely to apply to other groups. Perhaps the classic example is the seminal study of smoking and lung cancer in British doctors. Although carried out in a very unrepresentative sample of the British population (doctors on the medical register), it is clear that smoking is likely to cause lung cancer in other humans too!

For many environmental exposures the population attributable burdens can be large even when the relative risk is small (e.g. air pollution), but it is not always clear what the scope is for prevention or amelioration. For example, air pollution in London cannot be removed entirely and even significant reduction (by say 20 per cent) might entail quite dramatic technical, social and economic changes – which have their own costs and benefits. Moreover, in the short term, the improvement in health may not be as great as the attributable burden suggests, as there is likely to be some legacy from past exposures. Attribution is not the same as potential for prevention.

Finally, it is important to remember that the importance of an environment issue is not dictated by science alone. Indeed, in many cases, science may be subsidiary to issues of public perception and political context.

What are the implications?

The implications of a study very much depend on its nature and context. The following list provides headings that may need to be considered in relation to immediate practical steps, further investigation/surveillance and public policy. But of course, one study is very unlikely on its own to give rise to significant policy change, which should rather be based on careful review of the wider literature and balanced consideration of costs and effects.

- Immediate issues   public protection
  incident control/investigation team
  media/communication
- Further investigation immediate and longer term
  exposure studies
  risk assessment
  epidemiological study
  screening
  surveillance
- Policy   social and political context
  key players
  responsibilities
  consequences of actions
  legal vs. voluntary framework
  costs

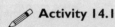

Activity 14.1

Read through the extract below from Keating et al. (2000) and write a discussion of its apparent findings and their implications for public health policy.

Heat-related mortality in warm and cold regions of Europe: observational study

Objectives: To assess heat-related mortalities in relation to climate within Europe.

Design: Observational population study.

Setting: North Finland, south Finland, Baden-Württemberg, Netherlands, London, north Italy, and Athens.

Subjects: People aged 65–74.

Main outcome measures: Mortalities at temperatures above, below, and within each region's temperature band of minimum mortality.

Results: Mortality was lowest at 14.3–17.3°C in north Finland but at 22.7–25.7°C in Athens. Overall the 3°C minimum mortality temperature bands were significantly higher in regions with higher than lower mean summer temperatures (P=0.027). This was not due to regional differences in wind speeds, humidity, or rain. As a result, regions with hot summers did not have significantly higher annual heat-related mortality per million population than cold regions at temperatures above these bands. Mean annual heat-related mortalities were 304 (95% confidence interval 126 to 482) in North Finland, 445 (59 to 831) in Athens, and 40 (13 to 68) in London. Cold related mortalities were 2457 (1130 to 3786), 2533 (965 to 4101), and 3129 (2319 to 3939) respectively.

Conclusions: Populations in Europe have adjusted successfully to mean summer temperatures ranging from 13.5°C to 24.1°C, and can be expected to adjust to global warming predicted for the next half century with little sustained increase in heat-related mortality. Active measures to accelerate adjustment to hot weather could minimise temporary rises in heat related mortality, and measures to maintain protection against cold in winter could permit substantial reductions in overall mortality as temperatures rise.

Feedback

This study compared heat-related mortality in seven cities/regions of Europe. Its apparent aim was to assess whether the burden of heat-related deaths was higher in warmer cities – a question which has bearing on the expected future burdens of temperature-related mortality under climate change.

What is not clear from the abstract alone, but can be guessed from your knowledge of how temperature effects are usually studied, is that the assessment of heat mortality was based on some form of time-series analysis. The main comparison is between the seven cities/regions, with high temperature as the exposure of interest, and the associated mortality as the principal outcome. Its primary conclusion is that the burden of heat deaths is no higher in warmer regions, indicating that populations have successfully adapted to their climatic conditions.

Underlying each of the city-level estimates of heat deaths is an analysis which entailed defining a 3°C-wide temperature band of lowest (daily) mortality, and then estimating how much mortality risk increased on hotter and colder days. As with any such analysis, you would wish to look at issues of control for time-varying confounders, model parameterization, time lags and autocorrelation, as described in Chapter 4 for air pollution. The details of this modelling can be important to the overall interpretation.

The authors provide rate estimates and confidence intervals for the number of deaths per million population attributable to heat in each area. These have been adjusted for other meteorological parameters. (Note, however, that the extract does not make it clear that the quoted numbers are in fact rates.)

The authors suggest that regions with hot summers did not have significantly higher annual heat-related mortality, but the abstract contains no global test for trend, though it does for the assertion that the minimum mortality temperature was higher in cities/regions with higher summer temperatures. Of course, with just seven regions, the evidence for or against a trend with mean seasonal temperature is bound to be limited.

When comparing the heat mortality in the different populations is it pertinent to consider whether like is being compared with like. Vulnerability to heat depends on population factors as well as on the temperature distribution. One factor of potential importance is population age, which should therefore be considered a potential confounder for comparisons between cities/regions. However, the analysis was confined to deaths in a fairly narrow age band of 65–74 years, so confounding by age is unlikely to be important. Other factors could be, however, including the underlying prevalence of cardio-respiratory disease.

The issue of cause and effect is somewhat complicated here as the purpose of the paper is to assess evidence for adaptation to high temperatures. The argument that people do adapt is certainly plausible, and the results are broadly consistent with the limited wider literature. But the general purpose of the paper is to demonstrate an absence of variation in heat vulnerability, which it does in limited degree.

Clearly vulnerability to temperature-related mortality is an important issue, as the large number of heat deaths in Europe in August 2003 demonstrate. It is partly reassuring that this paper provides some evidence that populations are able to adapt, though it may have limited relevance to the situation of rapid climate change predicted for coming decades, as adaptation is in part an issue of infrastructure and perhaps even genetic selection. It is not clear either how the results of this paper relate to the impact of extreme events. Clearly decisions about adaptation to climate change will be determined by a complex set of sociopolitical factors.

 Activity 14.2

Read the extract below from Wilkinson et al. (1999) and write a short discussion of its findings with particular reference to the interpretation of studies based on putative environmental hazards.

 Lympho-haematopoietic malignancy in relation to major oil refineries and associated industrial complexes

Objectives To examine the incidence of lympho-haematopoietic malignancy around industrial complexes that include major oil refineries in Great Britain after recent public and scientific concern of possible carcinogenic hazards of emissions from the petrochemical industry.

Design Small-area study of the incidence of lympho-haematopoietic malignancies, 1974–91, within 7.5 km of all 11 oil refineries (grouped into 7 sites) in Great Britain that were operational by the early 1970s and processed more than two million tonnes of crude oil in 1993.

Results Combined analysis of data from all 7 sites showed no significant (p<0.05) elevation in risk of these malignancies within 2 km or 7.5 km. Hodgkin's lymphoma, but no other malignancy, showed evidence (p=0.02) of a decline in risk with distance from refineries, while there was an apparent deficit of multiple myeloma cases near the refineries (p=0.04).

Conclusion There was no evidence of association between residence near oil refineries and leukaemias, or non-Hodgkin's lymphoma. A weak positive association was found between risk of Hodgkin's disease and proximity to major petrochemical industry, and a negative association with multiple myeloma, which may be chance findings within the context of multiple statistical testing.

↻ Feedback

This study examined lympho-haematopoietic malignancy around oil refineries and associated industrial complexes. It is a multi-site geographical (small area) analysis, with exposure defined by proximity to a refinery; the main outcome was incidence of lymphohaematopoietic malignancy as a whole and of its principal subgroups. The conclusion was that there was no clear evidence of association for the major disease subgroups, though there was weak evidence of increased risk near to refineries for Hodgkin's lymphoma.

As you will recall from the first three chapters of this book, understanding the context of this sort of investigation has an important bearing on its interpretation. If it were a single-site cluster investigation, then statistical inference would be difficult. But this is a multi-site study addressing a scientific hypothesis. The context is explained in the background — namely that it was undertaken following concerns about carcinogenic hazards from the petrochemical industry. You would wish to know whether those concerns were in part motivated by knowledge of the health statistics around any of the analysed sites, and if so, how that was dealt with in the analysis for this study. But assuming that the study was undertaken independently of the original concern, it can be treated as genuinely hypothesis-testing.

Confidence intervals are not shown but the extract reports no significant association with proximity to oil refinery sites for the main subgroups with the exception of Hodgkin's disease and multiple myeloma (for which proximity to a site is apparently protective!). Clearly, issues of multiple testing arise, and in part this lies behind the overall assessment that these apparently 'significant' results may be 'chance findings'. No mention was made of adjustment for confounding factors, but because the study was based on geographical analysis of routine data it is likely that few data were available on confounding factors other than age, sex and possibly socioeconomic status. Fortunately, there are few important confounders for most forms of lympho-haematopoietic malignancy, which generally have weak or absent associations with socioeconomic deprivation.

Because we are using area statistics, the inward migration of people from outside the locality will dilute the original 'exposed' population, and hence tend to weaken the apparent risk associated with the site (if there is one at all). However, lympho-haematopoietic malignancies have shorter time lag than solid tumours, so the potential bias is correspondingly slight. More problematic may be that proximity is a poor measure of exposure, and its use gives rise to a potential conservative bias.

From the extract we cannot comment on most of the Bradford Hill criteria. However, the possibility that the risk of Hodgkin's disease is increased by exposure to emissions from the petrochemical industry is plausible and does not conflict with other literature. The extract also implies that, for Hodgkin's, there was a decline in risk with distance from the site — which is a form of exposure-response gradient. But we are unable to judge the strength of association (no relative risks are shown), and the specificity of the result appears neither to strengthen or weaken the evidence in this case. The apparently protective effect in relation to multiple myeloma is counter-intuitive and adds weight to the interpretation of a chance finding.

Although the study related to a large study population and the diseases in question are serious, the underlying rate of Hodgkin's disease is fairly small, and so too therefore is the attributable number of cases. Given the uncertainty over the strength of evidence, the public health implications of this study appear slight.

Summary

You have learnt how the same general issues apply to the interpretation of studies of environment and health as to any epidemiological study. Consideration needs to be given to the influence of bias, confounding and random error. Causality can be assessed by reference to criteria such as temporal sequence, biological plausibility, the strength of association, specificity, consistency with other literature, biological gradient and coherence with the wider body of knowledge. In some circumstances, evidence may be available from natural experiments. In many cases, environmental exposures may give rise to small relative risks though attributable burdens can be large because of the ubiquity of exposure. But even where a large attributable burden is identified, the options for intervention need to be carefully considered in terms of the costs and benefits and the wider public health context.

References

Bradford Hill A (1965). The environment and disease: association or causation? *Journal of the Royal Society of Medicine* **58**: 295–300.

Keating WR *et al.* (2000) Health related mortality in warm and cold regions of Europe: observational Study. *BMJ* 321: 670–3.

Wilkinson P, Thakrar B *et al.* (1999). Lymphohaematopoietic malignancy around all industrial complexes that include major oil refineries in Great Britain. *Occupational and Environmental Medicine* 56(9): 577–80.

15 Emerging trends

Overview

In this final chapter you will look at the changing priorities in environmental health and consider some of the future directions for research. In high income countries, the primary focus of recent decades has been on local environmental exposures, with research based on traditional methods of epidemiology. In low and middle income countries, concern continues to focus on long-standing issues of poor access to clean water, sanitation and energy. However, as the climate change debate exemplifies, there is now a growing realization of the importance of global-scale changes including the effects of ecological disruption, biodiversity loss and depletion of natural resources, whose health impacts will present new challenges for the future. At the same time, particularly in the environmental field, there is recognition that methods need to be improved for studying subtle and potentially confounded effects of exposures. Methods in genetic epidemiology are likely to be applied with increasing frequency in this context.

Learning objectives

By the end of this chapter you should be able to:

- **describe the emergence of concerns over the health impacts of global environmental change**
- **be aware of some of the future research and policy needs**

Key terms

Adduct A chemical compound formed from the addition of two or more substances (e.g. forms of DNA after modification by chemical carcinogens).

Allele One of the (usually two) alternative forms of a gene.

Biomarker A cellular or molecular indicator of exposure, disease or susceptibility to disease.

Genome The genetic material of an organism.

Mendelian randomization Statement of the fact that inheritance of one trait is independent of other traits (except for associations over short segments of the genome).

Polymorphism The occurrence of a gene in several different forms.

Shifting focus of environment and health

There are many uncertainties in attempting to predict future directions of environment and health research. By definition, its new themes and methods have yet to be established, and experience shows that priorities can change rapidly as new evidence or new health threats emerge. However, a number of questions and research methods seem set to take an increasingly important role over coming years.

Changing priorities: from local to global change?

Over recent decades the primary environment and health concerns of the industrial world have been concentrated on chemical contamination of local environments. Pollution of the air, land and water has been the legacy of rapid and often poorly regulated industrial development. However, with growing wealth and improving environmental protection, many of these traditional environmental concerns have begun to assume less importance in high-income countries. For many environmental pollutants, there has been an inverted-U trajectory over time as exposures have risen and then fallen in parallel with increasing wealth and technological sophistication of societies. This applies, for example, to atmospheric concentrations of sulphur dioxide in high-income countries, to the release of heavy metals and to emissions from the nuclear industry. Falls have also occurred in bio-accumulating pesticides and other environmental contaminants, especially the long-lived ('residual') chlorinated hydrocarbons, which animal evidence suggests may be a hazard to immune, reproductive and neurological systems.

However, the relentless trends of urbanization and rapid industrial development in many middle- and lower-income countries has seen an increase in environmental degradation, often combined with hazards of poor sanitation, unsafe drinking water and urban poverty. On a global scale, environmental causes remain an important contributor to health burdens, as shown by the 2003 Global Burden of Disease Initiative (Table 15.1) (Ezzati *et al.* 2002). It is noteworthy that the dominating environmental impacts at global level remain, as they have done over many years, unsafe water, poor sanitation and inadequate access to clean energy (Figure 15.1). A consequence of lack of access to energy is the high prevalence of the use biomass fuels which, burned indoors, have detrimental impacts on respiratory health. These impacts are likely to continue to be major concerns for public health over the coming decades, especially in sub-Saharan Africa.

But a new focus of concern is global-scale environmental disruption. We recognize that human activity is producing profound changes to the earth's natural environment, reducing biodiversity (Millennium Ecosystem Assessment 2005), depleting non-replenishable resources and altering climatic conditions (IPCC 2001). Our impact on the composition of the atmosphere, with implications for global warming, was discussed in Chapters 11–13. But climate change is only one of a number of evolving large-scale environmental changes. The 2005 Millennium Ecosystem Assessment makes clear the scale of our impact on the natural world and the threats this carries for human health (see below). We are weakening many ecological systems, over-exploiting natural resources and contributing to an

Table 15.1 Global burden of disease: annual mortality and disability adjusted life years (DALYs) (millions)

	World	
	Mortality	DALYs
Childhood and maternal under-nutrition	6.16	227.5
Other nutrition-related risks and physical inactivity	18.8	183.9
Sexual and reproductive health risks	3.04	100.5
Addictive substances	6.92	128.9
Environmental		
Unsafe water, sanitation and hygiene	1.73	54.1
Urban outdoor air pollution	0.81	6.4
Indoor smoke from solid fuels	1.62	38.5
Lead	0.23	12.9
Global climate change	0.15	5.5
Occupational risks	0.78	24.6
Total mortality/dalys*	**55.9**	**1,455**

Total includes causes not separately listed

Source: adapted from Ezzati et al. (2002)

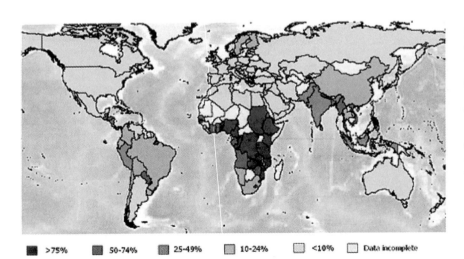

■ >75%   ■ 50-74%   ■ 25-49%   ▢ 10-24%   ▢ <10%   ▢ Data incomplete

Figure 15.1 Biomass as a per cent of national energy consumption
Source: ESRI (1996), DOE (2001)

unprecedented rate of loss of species. The long-term consequences of these changes for human health are unclear, but are likely to be negative (Aron and Patz 2001; McMichael 2001), and may represent 'a significant barrier to the achievement of the Millennium Development Goals (The World Bank Group 2005) to reduce poverty, hunger, and disease'.

Statement of the Millennium Ecosystem Assessment (2005) Living beyond our means: natural assets and human well-being

This statement was developed by the Board governing the MA process, whose membership includes representatives from U.N. organizations, governments through a number of international conventions, nongovernmental organizations, academia, business, and indigenous peoples.

The statement from the Board identifies 10 key messages and conclusions that can be drawn from the MA assessment (Millennium Ecosystem Assessment 2005):

* Everyone in the world depends on nature and ecosystem services to provide the conditions for a decent, healthy, and secure life.
* Humans have made unprecedented changes to ecosystems in recent decades to meet growing demands for food, fresh water, fiber, and energy.
* These changes have helped to improve the lives of billions, but at the same time they have weakened nature's ability to deliver other key services such as purification of air and water, protection from disasters, and the provision of medicines.
* Among the outstanding problems identified by this assessment are the dire state of many of the world's fish stocks; the intense vulnerability of the 2 billion people living in dry regions to the loss of ecosystem services, including water supply; and the growing threat to ecosystems from climate change and nutrient pollution.
* Human activities have taken the planet to the edge of a massive wave of species extinctions, further threatening our own well-being.
* The loss of services derived from ecosystems is a significant barrier to the achievement of the Millennium Development Goals to reduce poverty, hunger, and disease.
* The pressures on ecosystems will increase globally in coming decades unless human attitudes and actions change.
* Measures to conserve natural resources are more likely to succeed if local communities are given ownership of them, share the benefits, and are involved in decisions.
* Even today's technology and knowledge can reduce considerably the human impact on ecosystems. They are unlikely to be deployed fully, however, until ecosystem services cease to be perceived as free and limitless, and their full value is taken into account.
* Better protection of natural assets will require coordinated efforts across all sections of governments, businesses, and international institutions. The productivity of ecosystems depends on policy choices on investment, trade, subsidy, taxation, and regulation, among others.

Some of these changes may be associated with the emergence or re-emergence of infectious disease (Morens *et al.* 2004) or changes in infectious disease distribution (Table 15.2). Although not due to global environmental change, the recent example of West Nile Virus in the USA illustrates how rapidly new disease may become established (Campbell *et al.* 2002).

New methods?

The study of many environmental risks in future will require, and in large part be driven by, developments in research methods. You will recall from Chapter 4 that air pollution epidemiology has been through periods of comparatively high and low scientific interest as the perceived problems and the methods for investigating

Table 15.2 Examples of possible environmental links of infectious disease

Disease	Environmental link
Ebola haemorrhagic fever	Contact with reservoir
HIV/AIDS	Contact with primate reservoir
Hantavirus pulmonary syndrome	Increased rodent populations
Cyclosporiasis	Food importation
Malaria	Vector habitats
Influenza	Risks or benefits from increased human movements

Source: Aron and Patz (2001)

them have changed. Following the air pollution episodes in London, Donora and the Meuse Valley in the middle of the twentieth century, levels of air pollution rapidly fell in the UK as in other high-income countries and with it concern about the associated health effects. The debate about the health impacts of the lower levels of air pollution seen in cities today was rekindled by innovations in study design, specifically the time-series and semi-ecological cohort studies which enabled small relative risks to be quantified.

As we have repeatedly noted, many potential environmental hazards are likely to be associated with small relative risks which are difficult to disentangle from the effects of confounding factors. Because of the evident methodological challenges of quantifying such risks, there are those who argue that observational epidemiology is facing its limits (Taubes 1995). This conclusion may be premature, but it is clear that questions of the twenty-first century will pose new methodological challenges in environmental epidemiology.

One research method that offers promise, particularly in the field of cancer epidemiology, is the use of biomarkers. A biomarker is a cellular or molecular indicator of exposure to a hazardous agent, or of the disease risk associated with exposure. Perhaps the best known examples are the DNA adducts formed when chemical carcinogens react with and modify DNA. Modification of DNA by a chemical carcinogen is an early event in carcinogenesis, and hence the adducts may be of value in indicating exposure to the agent(s) in question or the risk that disease will develop (Shuker 2002). A variety of techniques of varying sensitivity and specificity are now available for measuring adducts of DNA or protein which may be used in molecular epidemiological studies (Farmer 1999). They may be used to help define biologically effective doses of specific chemical agents, and to inform risk assessments of low-level environmental exposures (Poirier 1997). Biomarkers have been studied for a range of hazardous substances, including polycyclic aromatic hydrocarbons (PAHs), aromatic amines, aflatoxins, nitrosamines and malondialdehyde.

Interest is also growing in use of genetic methods. Of particular relevance to environmental studies is the concept of 'Mendelian randomization'. The laws of Mendelian genetics mean that comparison of groups of individuals defined by *genotype* is equivalent to comparison of groups based on randomization, since genetic groups should not differ systematically (except for allelic associations over a short region of the genome) (Davey Smith and Ebrahim 2003) (Figure 15.2). This has led to the idea that Mendelian randomization may offer a way of investigating

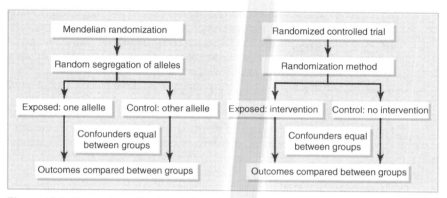

Figure 15.2 Comparison of design of Mendelian randomization studies and randomized controlled trials

Source: Smith and Ebrahim (2005)

environmental risks by using genetic polymorphisms that are known to influence exposures or have effects equivalent to those produced by modifiable exposures (Smith and Ebrahim 2005).

An example, cited by Smith and Ebrahim (2005) is that of the hazards of exposure to organophosphates in sheep dipping. Because workers generally know the possible effects of exposure, it is difficult to conduct an unbiased observational study. However, genetic variants determine the body's ability to detoxify organo-phosphates (through variation in the activity of the enzyme paraoxonase). Hence one would predict that, if organophosphates are a cause of ill health, symptoms would be greater in those (random) exposed individuals with lower paraoxonase activity. This has indeed been reported (Cherry *et al.* 2002) and provides evidence for symptomatic effects of organophosphates where more conventional epidemi-ological methods are difficult to interpet. As Clayton and McKeigue (2001) have observed, 'The ability of Mendelian randomisation to eliminate bias and residual confounding allows us to examine the effects associated with genetic polymor-phisms, even when these effects are small. Of special interest are polymorphisms that alter the metabolism of a dietary substrate or the activity of an enzyme or receptor'. The principal strength of genetic methods may not be in gene-environment interactions, but in testing specific causal pathways.

At the other end of the scale, new epidemiological methods will also have to meet the research needs for understanding the health consequences of broad environ-mental change. McMichael (2001) has argued that to understand 'the determinants of population health in terms that extend beyond proximate, individual-level risk factors (and their biological mediators), [we] must learn to apply a social-ecologic systems perspective'. He identifies the constraints with classical epidemiology as being: '1) a preoccupation with proximate risk factors; 2) a focus on individual-level versus population-level influences on health; 3) a typically modular (time-windowed) view of how individuals undergo changes in risk status (i.e., a life-stage vs. a life-course model of risk acquisition); and 4) the, as yet, unfamiliar challenge of scenario-based forecasting of health consequences of future, large-scale social and environmental changes'.

Concepts of life-course epidemiology (Lynch and Smith 2005) and of early-life influences on subsequent disease outcome (Barker 1995) are now well-established, but they have not yet featured large in the environmental field. Paying attention to life-course influences may be important when researchers wish to address signals of environmental risk against a background of epidemiological noise. But arguably the greater challenge is presented by the focus on global environmental change, which will require methods to understand the future, using evidence about the interplay of complex social, biological and environmental systems, and predictive models. This is an area of interdisciplinary work which remains unfamiliar to many epidemiologists. It is, however, one that seems set to assume increasing importance as we try to understand the potential threats of ecological disruption.

Summary

The chief focus of environmental epidemiology has changed over time. In many low-income countries, the dominant environmental concerns remain inadequate water, sanitation and access to clean energy. Trends of rapid industrialization and urbanization are also generating increasing burdens of environmental degradation in the developing world, while such concerns have begun to wane in the highest-income countries. However, we are now beginning to see the emergence of global-scale environmental change, which will present new challenges to epidemiology. The study of this will increasingly need to rely on innovations of research methods, which will also be needed to improve understanding of the confounded effects of other environmental hazards.

References

Aron J and Patz J (Eds) (2001). *Ecosytem Change and Public Health: A Global Perspective*. Baltimore and London, Johns Hopkins University Press.

Barker DJ (1995). Fetal origins of coronary heart disease. *BMJ* **311**(6998): 171–4.

Campbell GL, Marfin AA *et al.* (2002). 'West Nile virus.' *Lancet Infectious Diseases* **2**(9): 519–29.

Cherry N, Mackness M *et al.* (2002). Paraoxonase (PON1) polymorphisms in farmers attributing ill health to sheep dip. *Lancet* **359**(9308): 763–4.

Clayton D and McKeigue PM (2001). Epidemiological methods for studying genes and environmental factors in complex diseases. *Lancet* **358**(9290): 1356–60.

Department of Energy (2001). International energy annual 1999. Washington DC: DoE, Energy Information Administration.

Environmental Systems Research Institute (1996). World Countries 1995. Redlands, CA: ESRI.

Ezzati M, Lopez AD *et al.* (2002). Selected major risk factors and global and regional burden of disease. *Lancet* **360**(9343): 1347–60.

Farmer PB (1999). Studies using specific biomarkers for human exposure assessment to exogenous and endogenous chemical agents. *Mutat Res* **428**(1–2): 69–81.

IPCC (2001). *Climate Change 2001: The Scientific Basis. The Contribution of Working Group 1 to the Third Assessment Report of the Intergovernmental Panel on Climate Change*. Houghton JT *et al.* New York, CUP.

Lynch J and Smith GD (2005). A life course approach to chronic disease epidemiology. *Annual Review of Public Health* **26**: 1–35.

McMichael A (2001). *Human Frontiers, Environments and Disease. Past Patterns, Uncertain Futures*. Cambridge, Cambridge University Press.

Millennium Ecosystem Assessment (2005). *Millennium Ecosystem Assessment Synthesis Report.* Washington DC, Island Press.

Morens DM, Folkers GK *et al.* (2004). The challenge of emerging and re-emerging infectious diseases. *Nature* **430**(6996): 242–9.

Poirier MC (1997). DNA adducts as exposure biomarkers and indicators of cancer risk. *Environmental Health Perspectives* **105 Suppl 4**: 907–12.

Shuker DE (2002). The enemy at the gates? DNA adducts as biomarkers of exposure to exogenous and endogenous genotoxic agents. *Toxicology Letter* **134**(1–3): 51–6.

Smith GD and Ebrahim S (2003). Mendelian randomization: can genetic epidemiology contribute to understanding environmental determinants of disease? *International Journal of Epidemiology* **32**(1): 1–22.

Smith GD and Ebrahim S (2005). What can mendelian randomisation tell us about modifiable behavioural and environmental exposures? *BMJ* **330**(7499): 1076–9.

Taubes G (1995). Epidemiology faces its limits. *Science* **269**(5221): 164–9.

The World Bank Group (2005). *Millennium Development Goals.*

Useful websites

Institute for Environment and Health (UK): www.le.ac.uk/ieh/

Intergovernmental Panel on Climate Change: www.ipcc.ch/

Millennium Ecosystem Assessment: www.millenniumassessment.org/en/index.aspx

World Health Organization (Protection of the Human Environment): www.who.int/phe/en/

World Resources Institute, Washington: www.wri.org/

Appendix I
Clustering around a point source

I Introduction

A cluster of cases of disease is usually defined as any collection of cases which are unusually close in time, or space, or both. It is sometimes used more generally to define aggregation of cases by any dimension – location, season, time of day, age, race, religion, occupation . . . In this wider definition, study of the clustering of disease becomes a synonym for epidemiology. We shall assume the narrower definition.

We are mainly concerned here with studies to investigate a specific hypothesized cluster. The alternative, investigating clustering without a specific hypothesized cluster, is discussed briefly in the last section of these notes.

It is important to distinguish two contexts:

1 A causal hypothesis in search of a cluster
2 A cluster in search of a causal hypothesis

An example of the first would be a study of lung cancer in relation to proximity of residence to a coke-works (the clustering of cancer around a coke-works), instigated because animal experiments and occupational studies have shown that coke oven emissions may carry a risk of lung cancer. These kind of studies raise some methodological problems, mainly those common to all geographical epidemiology, but do not suffer from the major interpretational problems arising in the second context . . .

An example of the second context would be when a local newspaper or residents association perceives an unusually high number of cases of disease occurring in an area. A cause (a local factory, power lines, or similar) is usually then hypothesized. This is the context that is usually referred to as a *cluster report*. Interpretation of cluster reports is extremely problematic, and often controversial.

There may be confusion as to which of these contexts we are in. For the Sellafield leukaemia cluster, it is unclear which came first, the hypothesis (the nuclear reprocessing plant) or the cluster (a lot of leukaemias in one village). Usually, however, it is reasonably simple to identify at which end we are of the spectrum represented at the extremes 1/ and 2/.

Many of the statistical methods employed in contexts 1/ and 2/ are similar, but interpretation of their results should be very different.

2 Cluster reports

Stages of investigation

A cluster report will usually involve a group of cases (possibly covering a variety of diagnoses) in a particular area. If most of the cases are elderly, or if there is a variety of diagnoses, then a common cause is unlikely. The time of exposure and the latency of the disease are also important facts to consider. Reports of clusters are almost always associated with potential sources of toxic exposures. If the cluster is of a single and rare disease entity, and the relationship between exposure and the cluster is biologically plausible, then a further investigation is desirable.

In '*Guidelines for investigating clusters of health events*', published by the Centers for Disease Control, USA (Centers for Disease Control 1990), three stages in the further investigation of reported clusters are identified:

1   A preliminary evaluation to provide a quick rough estimate of the likelihood that an important excess of cases has occurred.
2   Case evaluation to verify diagnosis.
3   Occurrence evaluation to ensure all relevant cases have been recorded.

The process is one of screening. At each stage, it can be decided that there is insufficient evidence to warrant the resources required to go further. Most reported clusters lead nowhere.

The Texas sharp shooter

American humour has it that the 'Texas sharp shooter' (why Texas?) first fires his gun at the barn, then paints a target on the barn with the bull's-eye where the bullet-hole is. He then shows this off to his friends to show what a good shot he is. We should interpret cluster reports with something of the same caution that we should interpret the sharp shooter's claims of marksmanship. The local area in which the cluster is situated has come to your attention precisely because there was a lot of disease there. Disease occurrence rates will always vary randomly between areas. Has the 'target' been drawn around the area with high rates (there must be some), after the bullet has been fired?

The Texas sharp shooter logic is also known as *post-hoc, a posteriori*, or after-the-fact reasoning. If statistical significance tests are applied, it is equivalent to the multiple testing problem. Even if only one test is actually applied formally, implicitly many may have been carried out informally in the process that led to the selection of the area in question. Thus p-values obtained for clusters should be interpreted very cautiously. Some argue that they have no meaning at all.

Avoiding Texas sharp shooter logic

The Texas sharp shooter problem means that it remains very difficult to interpret a cluster that survives rigorous application of the DCD procedures. Statistical significance tests are useful to screen out apparent clusters which would in fact not be remarkable even if the area had been selected *a priori* – these will very rarely be worth following up. However, 'significant' results do very little towards proving that the cluster is 'real'.

If we cannot rely on significance tests, what can we do? Some possibilities:

1   Forget epidemiology – do an exposure and risk assessment.
2   Do an epidemiological study somewhere else with a similar exposure.
3   Look for things which distinguish the cases from others in the cluster area apart from residence there.

4   Investigate whether there is a dose-response relationship with exposure within the cluster area.

5   If a more exhaustive case ascertainment has been carried out, exclude the cases that were part of the original cluster before testing significance.

3 Example: Lung cancer in Vamdrup

The cluster concerned cases of lung cancer in the town of Vamdrup, in Denmark. Concern was focused around the rockwool plant. The data are unpublished and used with permission of Arne Paulson. A map of Vamdrup showing the location of cases is shown in Figure A1.1.

4 Comparing observed and expected numbers in the suspect area

The first step involves determining the appropriate geographic area and period in which to study the cluster. This can be difficult. The reported cluster will usually refer to highly selected geographic and time boundaries, which have been tightened to show the cluster in its most extreme form (one component of Texas sharp shooting). We should aim to use boundaries that correspond as much as possible to an area that could be said to be exposed to the putative hazard. It is often difficult to decide what this is. Further, to obtain expected numbers we need to use an administrative area for which population figures are known.

National or regional reference rates can be used to calculate the expected number of cases. This should preferably be done after stratification by age, using the method of indirect standardization. In this case the ratio of observed to expected cases is the SMR

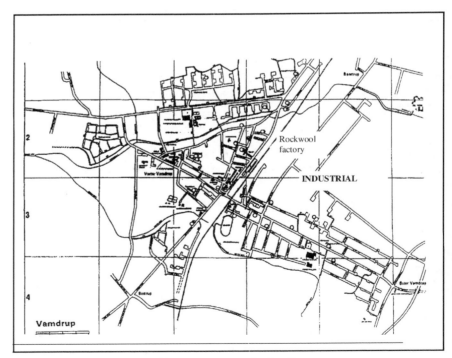

Figure A1.1 Vamdrup: cases of lung cancer are marked by ○

(standardized mortality ratio) for mortality or SRR (standardized registration ratio) for incidence. Otherwise, a crude calculation applying overall national rates to the overall population of the area would give a crude mortality or registration ratio.

Whether we obtain the expected number of cases allowing for age or not, an indication of chance uncertainty is obtained as follows: let D be the observed number and E the expected number for the chosen area and time period. Then D is an observation from a Poisson distribution with mean $\mu = \theta E$ where θ is a measure of the size of excess risk. If there is no excess risk $\theta = 1$. A one-sided p-value for $\theta = 1$ is obtained from the probability of observing D or more cases in a Poisson distribution with $\mu = E$ (or approximately from $z = 2(\sqrt{D} - \sqrt{E})$ using tables of the normal distribution). More informative than a significance test is the estimated value of θ, given by $\theta = D/E$. A 95 per cent confidence interval is given by the formula $(\sqrt{D} \pm 1.96/2)^2/E$. For a 90 per cent interval replace 1.96 by 1.645.

For Vamdrup the first convenient administrative area was the commune which contained Vamdrup. For this commune the observed number of cases was 39, during the period 1968–88, with expected number equal to 66.5. The SMR is $39/66.5 = 0.59$ or 59 per cent. As the observed number is less than the expected number, confidence intervals or significance tests are even less relevant than otherwise, however, for illustration: For a significance test, calculate

$$z = 2(\sqrt{D} - \sqrt{E}) = 2(\sqrt{39} - \sqrt{66.5}) = 3.82, \text{thus } P<0.001 \text{ (two-sided)}. -$$
a significant **deficit** of cases.

For 95% confidence limits calculate

$$(\sqrt{D} \pm 1.96/2)^2/E = (\sqrt{39} \pm 1.96/2)^2/66.5 = (0.42, 0.78)$$

Clearly there is no excess of cases for the commune overall, but 28 of the cases were in Vamdrup, so the low SMR would be more convincing if it referred just to Vamdrup.

5 Investigating a gradient of risk by exposure

If there is a measurement of exposure then it is important to see whether there is a relationship between risk and exposure. As we noted above, if we see this relationship within the reported cluster, this gives evidence in favour of a causal relationship independent of Texas sharp shooter logic. Unfortunately, however, with most clusters the measurement of exposure is very poor or missing altogether. Distance from a source of environmental pollution is sometimes used as a proxy for exposure.

The statistical technique for assessing the gradient of risk with exposure depends on whether the data are in the form of observed and expected by exposure, or as cases and controls by exposure.

Investigating a gradient of risk from observed and expected numbers

Calculation of expected numbers in areas defined by distance from the source of exposure require population counts in very small areas. As this is not available for the Vandrup example, we consider an example described in Hills (1992). Observed cases of cancer of the larynx and expected numbers from regional rates are shown below for bands between concentric circles drawn around the point source.

A plot of rate ratio (RR) against distance from the point source is shown in Figure A1.2. The appropriate statistical technique to estimate the trend of rate ratio with distance from the source is Poisson regression.

dist (km)	D	E	RR
0.50	0	0.27	0.00
1.0	0	1.03	0.00
2.0	10	7.38	1.36
3.0	10	6.93	1.44
4.9	24	15.90	1.51
6.3	12	10.41	1.15
7.4	9	8.28	1.09
8.3	11	8.58	1.28
9.2	8	9.97	0.80
10.0	6	8.58	0.70

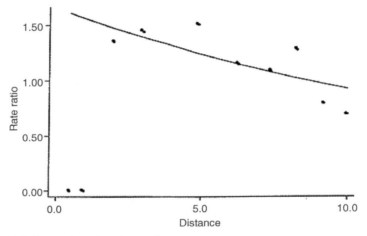

Figure A1.2 Rate ratio against distance from source

The stata command plus output is:

```
. poisson D dist, e(E) ir

Iteration 0: Log Likelihood = -20.214111
Iteration 1: Log Likelihood = -19.96582
Iteration 2: Log Likelihood = -19.965408


Poisson regression, normalized by E          Number of obs=10
Goodness-of-fit chi2(8)            =  6.357    Model chi2(1)=2.070
Prob > chi2                        =  0.6073   Prob > chi2=0.1502
Log Likelihood                     = -19.965   Pseudo R2 =0.0493


--------------------------------------------------------------------
   D  |     IRR     Std. Err.    z      P>|z|   [95%. Conf. Interval]
-------+------------------------------------------------------------
 dist |   .944141    .0375729  -1.444   0.149    .8732982   1.020731
--------------------------------------------------------------------
```

The rate ratio is estimated to be 0.944, which means that the rate changes by a factor of 0.944 for each increase of 1km in distance from the source. The change for 10 km is by a factor of $0.944^{10} = 0.562$, ie a decrease of roughly 50 per cent. The p-value is 0.149 so the factor is not significantly different from 1.

Investigating a gradient of risk from cases and controls

For Vamdrup the only measure of exposure was distance from the rockwool factory, and population figures were not available by distance from the factory, but only for the whole town. To investigate the relationship of risk with exposure in these circumstances it would be necessary to take a random sample of controls from people living in Vamdrup 10 or 15 years earlier, preferably stratified by age. This was not done, as the case was not considered to be strong enough, so as an illustration, I have chosen a random sample of control *locations* from the map of Vamdrup. This would not be a safe procedure in practice, because the locations might not correspond to houses, and anyway older people might live closer to the factory than younger people. However, it will serve as an illustration of techniques.

The question is whether the cases, on average, are closer to the factory than the controls. The statistical technique to estimate the trend in odds with distance is logistic regression. The stata command plus output is shown below.

```
. logistic fail dist

Logit Estimate                                  Number of obs =      56
                                                   chi2(1)    =   2.14
                                                Prob > chi2   = 0.0005

Log Likelihood = -32.744904            Pseudo ~2       = 0.1564

----------------------------------------------------------------------
    fail| Odds Ratio   Std. Err.      z    P>|z|    [95% Conf. Interval]
--------+-------------------------------------------------------------
    dist |  .0193509   .0257004   -2.970  0.003     .0014328   .2613388
----------------------------------------------------------------------
```

Here fail refers to the case/control status of each point and dist refers to the distance of each point from the source (S). The odds of being a case changes by a factor of 0.19 per km. The p-value for this is 0.003 showing the factor to be significantly different from 1.

6 Sellafield case study

Perhaps the most famous cluster of cases is the cluster of four cases of childhood leukaemia around the Sellafield nuclear plant, where only 0.25 were expected. The various stages in the investigation of this cluster have been well described by Gardner *et al.* (1990) and Gardner (1992).

7 Summary of problems in following up cluster reports

1 Choice of time period for incidence of cases, or mortality.
2 Choice of geographical region.
3 Choice of diagnostic group (and ascertainment of it).
4 Getting accurate population figures. Rapid changes of population over time (new

housing estates) and unusual aggregations of particular sorts of people in one place (military, elderly retired etc.) means that most clusters of disease are due to variations in population which have not been taken into account.

5   Measurement of exposure, particularly at the appropriate time.

6   Confounding effects of age and social class.

7   Interpretation of low p-values, in view of the Texas sharp shooter problem.

8 Investigating clustering without a specific hypothesized cluster

An investigation of clustering in space is part of the study of the geographical distribution of cases. The absence of any geographical pattern corresponds to a random distribution of cases with intensity depending only upon the population at risk. There are many possible departures from this. Smooth departures are usually called trends, departures where groups of cases tend to occur together are called clustering.

The methods for investigating clustering depend on how the information about the underlying population at risk is obtained, whether from population figures for small areas, or from a random choice of controls. The technicalities of these methods are beyond the scope of this course. Most share these features:

*   Their primary output is a significance test of the null hypothesis of no clustering. They usually also give an index of the extent of clustering, but these are often hard to interpret.
*   They do not tell you where the clusters are (or which clusters are real and which due to chance).
*   Some give an indication of the scale of clustering (clusters of the order of 1 km, 10 km, or 100 km).
*   Only a few can allow for confounding variables (which may induce clustering).

The usefulness of investigations of clustering is controversial. Rothman argues that all diseases are clustered, and that therefore it is of no interest to know that. Others argue that clustering suggests certain aetiologies, for example infectious agents. Perhaps the greatest usefulness of characterizing clustering is that it can then be taken into account in analysing the association of disease by a specific geographically distributed risk factor (especially in ecological correlation studies). If such background clustering is not allowed for, misleadingly low p-values may be obtained.

References

Centers for Disease Control (1990). Guidelines for investigating clusters of health events. *MMWR* **39**(RR–11): 1–23.

Gardner M (1992). Childhood leukaemia around Sellafield nuclear plant, in Elliott P, Cusick J, English D, Stern R Eds., *Geographical and Environmental Epidemiology*. Oxford, OUP: 291–309.

Gardner MJ, Snee MP *et al.* (1990). Results of case-control study of leukaemia and lymphoma among young people near Sellafield nuclear plant in West Cumbria. *BMJ* **300**(6722): 423–9.

Hills M (1992). Some comments on methods for investigating disease risk around a point source, in Elliott P, Cusick J, English D, Stern R, Eds., *Geographical and Environmental Epidemiology*. Oxford, OUP 231–7.

Appendix 2
Health guidelines for the use of wastewater in agriculture and aquaculture

Report of WHO Scientific Group, Geneva 1992.

Table A2.1 Survival times of selected excreted pathogens in soil and on crop surfaces at 20–300°C

Pathogen	Survival time	
	In soil	On crops
Viruses		
Enteroviruses*	<100 but usually <20 days	<60 but usually <15 days
Bacteria		
Faecal coliforms	<70 but usually <20 days	<30 but usually <15 days
Salmonella spp	<70 but usually <20 days	<30 but usually <15 days
Vibrio cholerae	<70 but usually <20 days	<5 but usually <5 days
Protozoa	<20 but usually <10 days	<10 but usually <2 days
Entamoeba histolytica cysts		
Helminths		<60 but usually <30 days<30 but
Ascaris lumbricoides eggs	Many months	usually <10 days<60 but usually
Hookworm larvae	<90 but usually <30 days	<30 days<60 but usually <30 days
Taenia saginata eggs	Many months	
Trichuris trichiura eggs	Many months	

Source: World Bank

* includes polio-, echo-, and coxsackie viruses

Measures of health risk from wastewater use

Knowledge of the survival patterns of excreted pathogens (Table A2.1) and of the removal of pathogens in wastewater treatment allows some assessment of the risk of the transmission of communicable diseases through wastewater use. This approach places greatest emphasis on microbiological criteria, and relies on pathogen removal to ensure the absence of 'potential' risks, but does not take account of the epidemiological concept of 'actual' or 'attributable' risk.

There is thought to be a potential risk – a risk (e.g. of developing a disease) that might, but does not at present exist – when pathogenic micro-organisms are detected in wastewater or on crops, even if no cases of disease caused by these micro-organisms are detected. This is in contrast to the epidemiologist's concept of risk, which focuses on the chance of an individual developing a given disease (or experiencing a change in health status) over a specified period as a result of a certain exposure. It is possible that a potential risk might not become an actual risk, because of factors related to pathogen survival, minimum infective dose, human behaviour, and host immunity.

In addition, a particular infection may have other routes of transmission in the community, so that some of the disease observed may not be associated with wastewater use. Risk is most usefully evaluated, therefore, on the basis of attributable risk or excess risk, which is a measure of the amount of disease associated with a particular transmission route within a population, in this case, the amount associated with wastewater reuse.

Measurement of attributable risk involves the comparison of two populations, one exposed to the risk factor of interest (in this case, wastewater use) and the other not so exposed (the 'control' population). Some cases of the disease of interest may occur in the control or unexposed population as the result of transmission via other routes (for example, diarrhoea transmitted through poor domestic water supplies and intestinal nematode infections transmitted through contamination of the domestic environment). The difference between the disease risk of the exposed and control populations – and not simply the amount of disease in the exposed population – is therefore a measure of the risk attributable to wastewater use.

The term 'relative risk' means the ratio of the risk estimates for the exposed and control populations and represents the number of times that disease is more (or less) likely to occur in the exposed as compared with the unexposed group. In the present case, it will provide a measure of the relative importance of wastewater reuse as a risk factor for the disease in question. However, in practice, it is probably more useful to assess the actual amount of disease caused by wastewater reuse, for which purpose attributable risk is the more convenient parameter.

The health risk associated with wastewater reuse may differ in different subgroups of the population. In this context, the most important subgroups to consider are persons consuming crops irrigated with the wastewater (consumer risk) and agricultural workers exposed occupationally (occupational risk). It is also important to consider persons of different ages separately, since the risk to children may be different from the risk to adults. The health protection measures to be taken will depend on whether consumer risks or occupational risks, or both, are to be minimized.

Factors involved in disease transmission

Many factors affect the degree to which the potential risk posed by a pathogen in wastewater can become an actual risk of disease transmission. For the agricultural or aquacultural use of excreta and wastewater to pose an actual risk to health, *all* of the following conditions must be satisfied:

1 *either* an infective dose of an excreted pathogen reaches the field or pond, *or* the pathogen multiplies in the field or pond to form an infective dose
2 the infective dose reaches a human host
3 the host becomes infected
4 the infection causes disease or further transmission

The risk is merely a *potential* risk if condition (4) is not satisfied. The agricultural or aquacultural use of excreta or wastewater will be of public health *importance* only if it causes an excess incidence or prevalence of disease or intensity of infection. Certain characteristics of a given pathogen will tend to increase the probable risk and public health importance of its transmission through wastewater reuse. These have been identified by Shuval *et al.* [1989] as follows:

- persistence for long periods in the environment
- long latent period or development stage
- low infective dose
- weak host immunity
- minimal concurrent transmission through other routes, such as food, water and poor personal or domestic hygiene

On this basis, the helminth infections in Categories III–V, caused by pathogens that are most persistent and have a long latent period and very low infectious doses, and to which host immunity is weak, can be expected to be among those posing the greatest actual risk from wastewater reuse. Where a significant amount of transmission occurs by other routes, as it often does with many of the faecal–oral infections (Categories I and II), a small amount of transmission due to wastewater reuse may be of relatively minor importance. The enteric virus diseases in Category II should be least effectively transmitted by wastewater use, despite the fact that they are moderately persistent and have low infectious doses. Concurrent transmission in the home is generally so intense that most infants acquire permanent immunity in the first years of life, so that there is little likelihood of additional exposure from wastewater reuse.

Current knowledge of the transmission of excreted pathogens thus suggests that helminth infection is the most important health risk and viruses the least important, with the bacterial and protozoal diseases falling between the two extremes. However, only epidemiological evidence can confirm the validity of this theoretical model.

Epidemiological evidence

Shuval et al. have rigorously reviewed all the available epidemiological studies on the agricultural use of wastewater. Their principal conclusions can be summarized as follows:

1 Crop irrigation with untreated wastewater causes significant excess infection with intestinal nematodes, where they are endemic, in both consumers and farm workers; the latter, especially if they work in the fields barefoot, are likely to have the more intense infections, particularly of hookworms
2 Crop irrigation with treated wastewater does not lead to excess intestinal nematode infection among field workers or consumers
3 Cholera, and probably also typhoid, can be effectively transmitted by the irrigation of vegetables with untreated wastewater
4 Cattle grazing on pasture irrigated with raw wastewater may become infected with 'Cysticercus bovis' (the larval stage of the beef tapeworm Taenia saginata); the actual risk of human infection is poorly documented but probably exists
5 There is only very limited evidence to show that, in communities with high standards of personal hygiene, the health of people living near fields irrigated with raw wastewater may be adversely affected either by direct contact with the soil, or indirectly through contact with farm labourers
6 Sprinkler irrigation with treated wastewater may promote the dispersion of small numbers of excreted viruses and bacteria in aerosols, but an actual risk of disease transmission by this route has not been detected

From the epidemiological studies it is clear that, when untreated, wastewater is used for crop irrigation, intestinal nematodes and bacteria present high actual risks, and viruses little or no actual risk (Table A2.2). The actual risks due to protozoa are not yet well established, as insufficient epidemiological data are available, but no studies have shown that wastewater reuse causes an additional risk of protozoal infection.

Table A2.2 Relative health risks from use of untreated excreta and wastewater in agriculture and aquaculture

Type of pathogen/infection	Excess frequency of infection or disease

Intestinal nematodes
  Ascaris spp
  Trichurls spp
  Hookworms
	High

Bacteria Bacterial diarrhoeas (e.g., cholera, typhoid)	Lower

Viruses
  Viral diarrhoeas
  Hepatitis A
	Lowest

Trematodes and cestodes
  Schistosomiasis
  Clonorchiasis
  Taeniasis
	From high to nil, depending upon method of excreta use and local circumstances

It has been suggested that the potential health risks associated with the aquacultural use of excreta and wastewater are threefold:

1 Passive transference of excreted pathogens by fish and cultured aquatic macrophytes
2 Transmission of trematodes whose life cycles involve fish and aquatic macrophytes (principally *Clonorchis sinensis* and *Fasciolopsis buski*)
3 Transmission of schistosomiasis

A review of the available epidemiological studies on excreta use in aquaculture found only one study in which the actual health risks associated with the passive transference of excreted pathogens were considered, but the results were inconclusive because of the epidemiological methodology employed. It found none dealing with occupational exposure leading to schistosomiasis. As far as trematode infections were concerned, they found that, while fertilization of ponds with excreta was important in the transmission of these diseases, so too was the faecal pollution of other local water bodies and ponds not deliberately fertilized with excreta. This is not an unexpected result, as the high degree of trematode multiplication in the snail host makes it possible for only slight and occasional contamination of surface water to give rise to relatively intense transmission.

Reference

Shuval HI, Wax Y, *et al.* (1989). Transmission of enteric disease associated with wastewater irrigation: a prospective epidemiological study. *American Journal of Public Health* **79**(7): 850–2.

Appendix 3
Epidemiological formulae

Risks, rates and risk/rate ratios

d is the number of observed (new) cases of disease

N is the number of people at risk (disease free at the start of follow up)

t_i if the follow up time (in person years) of the ith person

Risk = d/N

Rate = $d/\Sigma t_i$

$$\text{Risk (or rate) ratio (RR)} = \frac{\text{risk (or rate) in exposed}}{\text{risk (or rate) in unexposed}}$$

Odds and odds ratios (case control studies)

$$\text{Odds (of exposure)} = \frac{\text{number (proportion) exposed}}{\text{number (proportion) unexposed}}$$

$$\text{Odds ratio (OR)} = \frac{\text{odds of exposure in those with disease}}{\text{odds of exposure in those without disease}}$$

$$= a/c \ / \ b/d = ad/bc$$

	Disease	No disease	Total
Exposed	a	b	a+b
Not exposed	c	d	c+d
Total	a+c	b+d	N=a+b+c+d

Population attributable risk (PAR) and population attributable risk fraction (PAF)

r_0 is the risk (or rate) in the unexposed group

r_1 is the risk (or rate) in the exposed group

r is the risk (or rate) in the total study population

p is the proportion of exposed in the population

RR is the risk ratio (rate ratio, odds ratio)

$PAR = r - r_0$
or
$PAR = p\ (r_1 - r_0)$

$PAF = PAR/r$
so
$PAF = (r - r_0)\ /\ r$
or
$PAF = p\ (RR-1)\ /\ [\{p(RR-1)\}+1]$

Standardized mortality (or morbidity) ratio (SMR)

O is the observed number of deaths (cases)

E is the expected number of deaths (cases)

$$SMR = \frac{O}{E} \times 100$$

Expected deaths are the number of deaths that would be expected if age-, sex- and calendar period-specific rates for the general population applied in the cohort: $E = \Sigma R_i n_i$, where R_i is the rate of death (disease) in stratum i in the reference population and n_i the population at risk in the stratum i of the study population.

Glossary

Adaptation Strategies, policies and measures undertaken now and in the future to reduce potential adverse impacts of climate change.

Adaptive capacity The general ability of institutions, systems and individuals to adjust to potential damages, to take advantage of opportunities or to cope with the consequences of climate change in the future.

Adduct A chemical compound formed from the addition of two or more substances (e.g. forms of DNA after modification by chemical carcinogens).

Aerobic Living or taking place in the presence of air or oxygen.

Allele One of the (usually two) alternative forms of a gene.

Anaerobic Living or taking place without air or oxygen.

Becquerel (Bq) A unit of radioactivity. Specifically, one Bq is the amount of the radioactive material that will have one disintegration in one second.

Biogas Gas consisting mainly of methane produced by anaerobic digestion of organic waste.

Biomarker A cellular or molecular indicator of exposure, disease or susceptibility to disease.

Chromosome Structure(s) found in the nucleus of a cell, made of DNA and proteins, that contains genes. Chromosomes usually come in pairs.

Climate The average state of the atmosphere and the underlying land or water in a specific region over a specific time scale. Should be distinguished from 'weather', which is the atmospheric conditions in a specific place at a specific time.

Climate change A statistically significant variation in either the mean state of the climate or in its measurable variability, persisting for an extended period (typically decades or longer).

Climate change in mitigation An anthropogenic (human) intervention to reduce the sources or enhance the sinks of greenhouse gases.

Climate variability Variability in the mean state and other statistics (such as standard deviations, the occurrence of extremes etc.) of the climate on all temporal and spatial scales beyond that of individual weather events.

Coliforms A group of bacteria, some of which (faecal coliforms), are normally found in human and animal faeces.

Congenital anomaly (malformation) Developmental defects present at birth.

Critical period for congenital anomalies The stage of development of an embryo when it is most susceptible to teratogenic effects. Differs for different organs/organ systems.

Cumulative exposure The total of exposure summed over time, usually the multiplication of the level of exposure (for each job an individual has held) by the duration exposure, summed over all jobs/time periods.

Disability adjusted life year (DALY) A measure of health based not only on the length of a person's life but also their level of ability (or disability).

Disease cluster An unusual aggregation of health events that are grouped in space and time.

Dose response The magnitude of the effect of a given level of exposure to an agent.

Effluent Outflowing liquid.

Electromagnetic spectrum Radiation, including visible light, radio waves, gamma rays, x-rays, in which electric and magnetic fields vary simultaneously.

Embryological (developmental) window The time period in which a foetus or embryo is most vulnerable to exposure to a teratogenic agent, after which there is less risk of inducing major congenital anomalies.

Extreme weather events Events that are rare within their statistical reference distribution at a particular place.

Genetic effects Effects seen in the offspring of an exposed individual (parent or grandparent) rather than in the individual themselves as a result of damage to genetic material. For an effect to be genetic, exposure must be before conception.

Genome The genetic material of an organism.

Genotype The individual genetic make-up which may affect susceptibility to a teratogenic agent.

Geographic information system An information system used to store, view and analyse geographical information.

Gray (Gy) The absorbed dose of radiation corresponding to one joule per kilogram.

Infective dose The number of pathogens which must simultaneously enter the body, on average, to cause infection.

Ionizing radiation Radiation that is sufficiently energetic to break the bonds that hold molecules together to form ions.

IPCC Intergovernmental Panel on Climate Change.

Job-exposure matrix A list of job titles each with an estimated exposure linked to it. Typically, exposure measurements are not made of all workers but rather of a sample of workers which are then applied to other workers with the same job titles.

Mendelian randomization Statement of the fact that inheritance of one trait is independent of other traits (except for associations over short segments of the genome).

Mortality displacement (harvesting) The name given to the bringing forward in time by just a few days or weeks of death or other health event by an environmental exposure.

Mutagen An agent which can cause genetic damage to individual cells.

Non-ionizing radiation Radiation which does not cause the disruption of molecular bonds and hence does not form ions.

Particulates Particulate matter, aerosols or fine particles of solid or liquid suspended in the air.

Phenotype The observable characteristics of the individual.

Polymorphism The occurrence of a gene in several different forms.

Post hoc hypothesis Formulation of hypothesis after making the observation.

Raster A form of spatial data representation in which the data are stored as a matrix of cells or pixels.

Residual confounding Distortion of the exposure-effect relationship (confounding) that remains after attempted adjustment for the effect of confounding factors.

Scenario A description of a set of conditions, either now or, plausibly, in the future.

Semi-ecological design A term often applied to cohorts studies of air pollution impacts on health in which exposure is defined at group level (by centrally-located pollution monitor) but data on other risk factors are available at individual level.

Sewage Human excreta (faeces and urine) and wastewater, flushed along a sewer pipe.

Sievert (Sv) A unit of equivalent dose of radiation which relates the absorbed dose in human tissue to the effective biological damage of the radiation. A milisievert (mSv) is one thousandth of a sievert.

Sullage Domestic dirty water not containing excreta, also called grey water.

Teratogen An agent which can induce congenital anomalies in a developing foetus.

Teratogenic effects Abnormalities in the embryo or foetus produced by disturbing maternal homeostasis or by acting directly on the foetus *in utero*.

Texas sharp shooter phenomenon A term used to refer to *post hoc* studies: the Texas sharp shooter shoots first, then draws the target where most bullets have hit. The epidemiological analogy is the selection of a cluster from the pool of all potential clusters.

Time-series studies The analysis of variation in events, such as daily or weekly counts of deaths or hospital admissions, in relation to exposures measured at similar temporal resolution.

Vector (in mathematics and physics) A quantity having both direction and magnitude which determines the position of one point in space relative to another.

Vector An organism, such as an insect, that transmits a pathogen from one host to another.

Vector-borne diseases Diseases that are transmitted between hosts by a vector organism such as a mosquito or tick – (e.g. malaria, dengue fever, leishmaniasis).

Vectorial capacity The average number of potentially infective bites of all vectors feeding upon one host in one day, or, the number of new inoculations with a vector-borne disease transmitted by one vector species from one infective host in one day.

Vulnerability The degree to which individuals and systems are susceptible to or unable to cope with the adverse effects of climate change, including climate variability and extremes.

Water scarcity Not enough water to supply all users' needs.

Water security A situation of reliable and secure access to water over time. It does not equate to constant quantity of supply as much as predictability, which enables measures to be taken in times of scarcity to avoid stress.

Water shortage A situation where levels of available water do not meet defined minimum requirements.

Water stress The symptomatic consequence of scarcity which may manifest itself as decline in service levels, crop failure, food insecurity etc. This term is analogous to the common use of the term 'drought'.

Index